# SHALE OIL

# SHALE OIL

## Tapping the Treasure

## Robert Alden Loucks

Library of Congress Number:     2002091921
ISBN:        Hardcover     1-4010-5703-9
           Softcover     1-4010-5702-0

The author and publisher has made every effort to ensure the accuracy and completeness of information contained in this book. Any slights of people, places, or organizations are unintentional.

First printing 2002.

ATTENTION CORPORATIONS, UNIVERSITIES, COLLEGES, AND PROFESSIONAL ORGANIZATIONS: Quantity discounts are available on bulk purchases of this book for educational or gift purposes. For information, please contact Xlibris.

Cover Photograph from C-b Project files and given to me by my good friend and fellow oil shaler, Bob Thomason. This book was printed in the United States of America.

To order additional copies of this book, contact:
Xlibris Corporation
1-888-795-4274
www.Xlibris.com
Orders@Xlibris.com
14886

# CONTENTS

## DEDICATION FOR SHALE OIL— TAPPING THE TREASURE

This book is dedicated to my eight wonderful grandchildren as well as their parents and grandmother. I sincerely wish that the United States in which they grow up and live will be as great a place for them as it has been for me. My hope is that this work will stir interest in utilizing our natural shale oil resource, which I feel will be a significant contributor to the future well-being of our country.

*I have placed my rainbow in the clouds*
*as a sign of my promise*
*until the end of time.*

*Genesis 9:13*

# PREFACE

OUR COPY OF the Rand McNally Road Atlas has an introductory section entitled "America's Love Affair with the Road", which fittingly summarizes steps leading to our dependence upon our motor vehicles. Effects of advertising (Burma Shave), world wars, and especially government (National Parks, the Interstate system) on the growth of automobile usage are shown. The last entry provides the gee-whiz statistics, "More than 175 million licensed drivers in the U.S.; 200 million registered vehicles: 4 million miles of paved roads and highways."

I might expand the history to include other motor driven vehicles. During a trip to Lake Powell, it became even more evident to me just how much we love and count on hydrocarbon fuels to live and have fun. Four of us were going fishing at the lake, in two large, powerful boats. To preserve the independence of our schedules, we drove to the lake, about 150 miles distant, in three different large vehicles. I arrived a little before the others and sat by the boat ramp and watched people. Boat trailers were lined up to either launch or take out boats of all sizes and styles. The trailers were pulled by everything from bus-sized mobile homes to smallish sedans. I saw power plants ranging from air-cooled two cycle engines to massive diesels. Even out in the wide-open spaces surrounding Lake Powell, the noise and aroma were significant.

A while later I paid a call on my granddaughter who was then working in a recreational vehicle store in Yuma, Arizona. Once again, I was impressed by the imagination of entrepreneurs. Available were vehicles for land or water transportation designed to excite nearly everyone's interest. Motorcycles, four-wheeled ATVs as powerful as many road cars, jet skis capable of extremely high speeds carrying up to four people and all the accessories were immediately available for our affluent consuming society.

This love affair of USers with motorized vehicles has been a particular concern of mine since I was fortunate enough to be involved with the petroleum business for most of my adult life. (Note: I have long been bothered with the proper designation for residents of the United States of America. "Americans" has always seemed improper because that designation should include all South, Central and North Americans. "Yanks" or "Yankees" may exclude southerners, or National League baseball teams. I'm just not sure about "Amis". I've settled on "USers"—pronounced "you-essers"—because it alludes to our well-earned, but materialistic, disproportionate exploitation of world resources.) The oil industry was an exciting business in which to be. I was involved with the design and construction of a number of fuel-producing facilities for Shell Oil Company in the 50's, 60's and 70's. In 1975, I was fortuitously assigned to an oil shale project in Colorado. The economics of oil shale development soon became suspect to careful energy company management and I found myself working on the same project but for a different company. Occidental Petroleum, headed by the visionary Dr. Armand Hammer, assumed Shell's obligations for the shale oil project and, in the process, I did something I never expected to do and left Shell for another company.

Now enjoying retirement, I have begun to clear out accumulations of my working files. Reading through the drawers of meeting notes, contracts, personnel records, etc., from the oil shale days, it seemed more and more compelling to me to somehow record for posterity some of the excitement that prevailed during the decade from 1975 to 1985 here in western Colorado. It was a singular experience for all of the participants. Utilizing a habit inculcated by Ron Brown, an early hero of mine in Shell, I have retained detailed handwritten notes of most of my involvement in anything to do with my job. That, plus my good fortune to be near the center of much of the action occurring in the saga of oil shale development allows me to, perhaps uniquely, describe what happened.

Some of my inspiration for recording this history comes from the cover photograph. Shale oil development, as the promised pot of gold at the end of the rainbow, could very well be an important part of the answer to our country's energy problems. Herein is expressed my long-held conviction that ecological concerns about shale oil production expressed by slow- or no-growth activists have been thoroughly researched and acceptable answers provided. Development of this vast native resource should proceed on a measured and methodical schedule to allow today's young people to enjoy the benefits of a strong United States as those of my generation were so blessed.

# CHAPTER 1

# The Beginning

THE FIRST MATERIAL in my shale oil files is dated May 8, 1974. I had recently been reassigned to Shell Oil's head office engineering staff in Houston after what seemed to me to be an interminable assignment in refinery operations. It's always been clear to me that I thrive on project management rather than operational work. It is second nature for me to succeed if I can see a deadline to meet, rather than facing day-to-day routine. I heard on Public Radio the other day something that I wished I had learned a long time ago. The commentator called those who work in the Intensive Care Unit of a hospital, "intensivists". If I'd known that, I believe that I should have carried the title of Projecteer or the like. The Projecteer motto, as passed on to me by another of my Shell hero's, Bob Heinze, is, "Just tell me what you want me to do and then get out of the way."

As part of the new head office assignment, I was given a special job to represent our part of the company in Shell's ongoing effort to investigate and potentially develop a shale oil production capacity. Shell had been active in the shale oil business for several years. Shell research teams had been participating since the 1950's in industry-wide mining and retorting investigations as well as a unique Shell-only effort to recover shale oil *in-situ* using superheated steam.

In the 1970's however, there seemed to be an intensive, vibrant movement afoot to develop a discrete and new industry dedicated to production of shale oil as a replacement for diminished national oil production. This effort was stimulated by the recent oil embargo, lines at gasoline stations and other cultural shocks which surprised USers and much of the rest of the world into a realization of our dependence on petroleum and the fact that much of it came to us from strange places.

Interest in oil shale had risen and fallen over the years. The increase in interest was usually associated with a war or similar crisis, which highlighted the fact that curtailment of overseas shipping could seriously impact industry and the military. The lack of a locally controlled source of petroleum energy is always viewed as a crisis situation and the supply of US petroleum is known to be a depleting resource. This is always a source of confusion to all, because the powers that be are always quoting a US reserve of so many billions of barrels of oil which when divided by the daily usage rates shows that we'll run out some time in the not too distant future. However, when next looked at, it appears that even though usage rates may have even increased, the reserves are still about where they were. This is at least partly because that oil exploration is always taking place at greater or lesser levels dependent on the current price of oil. As a result, new deposits are continually being added to the reserves to make up for some of those depleted. Others also suggest that reserves for OPEC nations have been arbitrarily re-calculated and raised when the organization decided to allow daily production based on reserves as well as previous output.

Despite this phenomenon of oil production statistics/economics, the fact remains that petroleum is a depleting resource. In a book copyrighted in 1969 by the National Academy of Sciences, "Resources and Man"[i], M. King Hubbert then working in the U.S. Department of Interior provided the following conclusions concerning the world's supply of fossil fuels:

" . . . the principal results of the (provided) estimates of the approximate magnitudes of both the United States' and the world's supply of the fossil fuels are the following:

If these substances continue to be used principally for their energy contents, and if they continue to supply the bulk of the world's energy requirements, the time required to exhaust the middle 80 percent of the ultimate resources of the members of the petroleum family—crude oil, natural gas, and natural-gas liquids, tar-sand oil, and shale oil—will probably be only about a century.

Under similar conditions, the time required to exhaust the middle 80 per-cent of the world's coal resources would be about 300 to 400 years (but only 100 to 200 years if coal is used as the main energy source).

To appreciate the bearing of these conclusions on the long-range outlook for human institutions, the historical epoch of the exploitation of the world's supply of fossil fuels is shown graphically in Figure 8.27, where the rate of production of the fossil fuels as a function of time is plotted on a time scale extending from 5,000 years ago to 5,000 years in the future—a period well within the prospective span of human history. **On such a time scale, it is seen that the epoch of the fossil fuels can only be a transitory and ephemeral event—an event, nonetheless, which has exercised the most drastic influence experienced by the human species during its entire biological history.**" (My emphasis added).

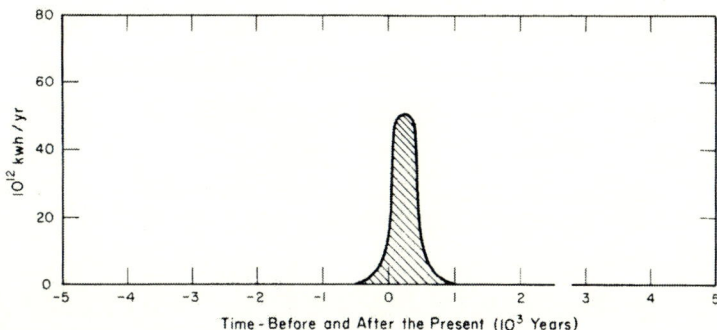

**FIGURE 8.27**
Epoch of exploitation of fossil fuels in historical perspective from minus to plus 5,000 years from present. (From Hubbert, 1962, Figure 54, p. 91.)

It is only a matter of when, not if, fossil fuels will all be used up and other energy sources will be required. Petroleum, as well as other energy sources, is a depleting resource. Mr. Hubbert, as he developed his thesis and provided more and more objective data, predicted that US domestic production of oil would peak around 1970 and would require increasing amounts of foreign oil to supply United States ever-increasing demands. And, true to his startling prediction, beginning in the 1970's and continuing on to the present, the amount of oil being imported into the US from outside our borders is a significant percentage of the daily amount used.

(In a recent book[ii], Hubbert's contemporary, Kenneth Deffeyes carries the premise forward to a prediction for the entire world supply of hydrocarbon. Using the same system and same kind of data available to Hubbert, Deffeyes suggests that the peak of the Hubbert Curve for all petroleum could be as soon as 2004. Given that Hubbert was not given serious attention until his prediction proved true in 1970, it is probably worthwhile to pay attention to these current worldwide forecasts.)

As a direct result of these circumstances and as a reaction to militant actions by world oil producers, oil shortages, real and perceived, accompanied by a dramatic (and real) increase in the world price of oil resulted in 1973. An intensified effort was initiated to develop an oil shale production capacity to reduce our reliance on outside-our-border oil and protect our industrial and military capabilities. As noted, this campaign was but the latest of a series of such attempts over the years, each a reaction to possible or real scarcities of world oil supplies.

The significant amount of exploration and development work previously performed had established without question that about two trillion (that's 2 followed by 12 zeros) barrels of hydrocarbon exist in the tri-state area of northwestern Colorado, northeastern Utah and southwestern Wyoming. This amount of oil is greater than established reserves in Saudi Arabia, and over 100 times larger than the petroleum deposits beneath the currently controversial ANWR, Alaska's Arctic National Wildlife Refuge. Utilization of this vast amount of domestic energy has always appeared to be an exceptional opportunity for speculators and entrepreneurs. Pioneer developers have repeatedly shown, however, that producing oil from rock is more difficult and more expensive than it would appear. Shale oil production economics have never compared favorably to those of conventional drilled oil well costs.

Oil shale as a energy source had been known of for years as described in the following oil shale histories: "The Rock that Burns" (1967) by Harry K Savage; "The Elusive Bonanza; The Story of Oil Shale—America's Richest and Most Neglected Natural Resource" (1970) by Chris Welles; "History of Western Oil Shale" (1980) by Paul L. Russell; and "World Oil Shale Deposits" by Charles O. Hook and Paul L. Russell in the January 1982 issue of *Mining Engineering*. A brief summary of the history would include the established beneficial use of shale oil in Scotland, Eastern Europe and other places; the development of an understanding of the vast amount of hydrocarbon-type energy from oil shale available in the United States; and recognition that the most likely US deposits for large scale oil shale development existed in the three states of Colorado, Utah and Wyoming. Hook and Russell scrupulously described the world reserves of oil shale as follows:

"Kerogen is the organic constituent of oil shale and may be the most abundant form of hydrocarbon on earth, **possibly exceeding coal occurrences and far more abundant than naturally occurring petroleum resources.** (My added emphasis). Oil shale, by one name or another, is widespread and is known to exist in more than 38 countries.

## Table 1 - World Oil Shale Occurrences

**Africa:** Morocco, South Africa, Zaire
**Asia:** Burma, China, Israel, Jordan, Siberia, Syria, Thailand, Turkey
**Australia**
**Europe:** Austria, Bulgaria, Czechoslovakia, France, Germany, Italy, Luxembourg, Portugal, Romania, Scotland, Sicily, Spain, Sweden, Switzerland
**(Former) USSR:** Yugoslavia
**New Zealand**
**North America:** Canada, Costa Rica, United States
**South America:** Argentina, Brazil, Chile, Paraguay, Uruguay

Oil shale industries have existed in at least 14 countries at one time or another. Many predated the Drake oil discovery in Pennsylvania in 1859, and most were unable to compete with natural oil; these operations either converted to other uses or closed. At the present time (1982) shale oil is being produced on a commercial scale only in the former USSR and in China.

Only very rough estimates of grade and quantity are available for most of the worldwide shale occurrences since only a few deposits have been thoroughly explored. The Green River Formation of Colorado, Utah and Wyoming is an exception. For many years this oil shale deposit has been described as the largest single concentration of hydrocarbon material in the world. With increasing world interest, larger deposits may yet be found.

In 1965, the US Geological Survey published an in-place world oil shale estimate. This study, while made some 16 years ago, is still a reasonable order-of-magnitude estimate. (Still reasonable in 2002 in my opinion). Although exploration and evaluation are underway and a few oil shale deposits have been appraised, information on which to evaluate most worldwide oil shale occurrences has improved little since 1965. (Also still valid).

Table 2 presents an order-of-magnitude estimate of in-place world oil shale resources. Total US resources and Green River Formation resources have been added to the USGS table for comparison purposes. On a world scale, the oil shales of the Green River Formation can be considered about average grade. However, their thickness and consistency of grade make them the world's largest known deposit.

# Table 2 - In-Place Shale Oil Resources
## Order-of-Magnitude Estimates

| Location | Area Underlain by Organic Shales (Square Miles) | Oil Equivalent (Trillions | in of | Deposits Barrels) |
|---|---|---|---|---|
| | | 25 gal/st to 100 gal/st | 10 gal/st to 100 gal/st | 5 gal/st to 100 gal/st |
| Africa | 5 million | 4 | 84 | 534 |
| Asia | 7 million | 5.5 | 115 | 705 |
| Australia | 1.2 million | 1 | 21 | 121 |
| Europe | 1.6 million | 1.5 | 27 | 167 |
| North America | 3 million | 3 | 53 | 313 |
| South America | 2.4 million | 2 | 42 | 252 |
| World Total | 20.2 million | 17 | 342 | 2092 |
| United States | 1.6 million | 2 | 28 | 168 |
| Green River | 16.5 thousand | 1.2 | 4 | 8 |

gal/st =gallons of oil per short ton of shale rock

The numbers in Table 2 are subject to debate, but they do present a basis for some observations. For example, world oil shale oil resources are huge in comparison to world petroleum reserves, even if grades of only 25 gal/st and above are considered. If oil shale grades down to 5 gal/st are included, the resource estimate becomes astronomical. The key question is, of course, how much is actually recoverable? Based upon today's technology and economics, only a small percentage can be considered recoverable, and shale as low grade as 5 gal/st may never be recovered." (Still valid in 2002).

Hook and Russell go on to describe shales of interest in the US.

"It has been estimated that an equivalent of 7 trillion barrels of oil is contained in US oil shale reserves. Nearly 60% is in the relatively rich Green River Formation of Colorado, Utah, and Wyoming. The remainder, for the most part, is contained in the leaner Devonian-Mississippian black shales that underlie much of the eastern part of the US. For a variety of technical and economic reasons, only a fraction of the in-place reserve could ever be exploited, but even this fraction could provide a significant addition to conventional energy supplies . . .

The Devonian-Mississippian shales extend over some 400,000 square miles and may contain as much as 3 trillion barrels of oil.

These shales are thinner and lower grade than Green River shales and their economic value has yet to be established. It is estimated that (a substantial portion) can be surface mined.

Other oil shales of interest include:

**Alaskan marine shales**, of Triassic and Jurassic age, in several areas of Alaska. Grade and reserves are not determined. However, upper Triassic shales, adjacent to the National River may yield about 30 gal/st and contain about 2 million barrels. Unexplored extensions of similar grade shales are estimated to be about 200 billion barrels. Black marine shales of northern Alaska are extensive and these may contain 250 billion barrels in over 25 gal/st oil shales.

**Heath Formation, Montana.** These formations could yield oil along with vanadium, molybdenum, nickel, zinc, and selenium. Shale oil potential may be more than 180 billion barrels with oil yield of 10 gal/st.

**The Woodruff Formation, Nevada** contains 0.8%vanadium pentoxide in low to moderate grade oil shale.

Oil shales may also occur in the northern part of the Appalachian Basin, the Phosphoria Formation of Idaho, Montana, Wyoming, and Utah; and underlying the Great Plains and Rocky Mountains."

Despite these enormous reserves of potential hydrocarbon fuels and raw materials for plastics and other synthetics, previous attempts to develop an industry failed for various reasons. Savage suggests a misplaced emphasis on process engineering and a lack of attention to mining (materials handling) technology. Welles suggests that oil companies had, and have, a strong reluctance, if not aversion, to development of an alternate energy industry because of the fear of establishing competitors which may be subject to different regulations and hence have an unfair advantage over the well established oil production system.

"He (Welles)became convinced the previous granting of patents was not due to legal misinterpretation or lax administration but to a massive conspiracy of rapacious oil companies, unprincipled speculators, and traitorous Interior employees." P154.

Also Welles commenting on Interior's position as stated in BLM Colorado Contest #359 (*U.S. v. Frank W Winegar*) "these men who actually set up producing shale oil retort (*sic*) were 'irrational gamblers' who were 'mesmerized' by the speculative fever. No Prudent Man would possibly have become involved." P167.

Notwithstanding all these considerations, I enthusiastically joined a long line of adventurous souls attempting one more time to establish a viable and profitable shale oil industry in the United States. Welles' Prudent Man concept was about to be tested once again. Further, I was searching for some sort of mystical intercession such as that signified by the promise of the rainbow in Genesis and/or the pot of gold which I've been told all of my life exists at the end of a rainbow.

# CHAPTER 2

# The Energy Crisis

HOW DID THIS 1973 effort to develop US shale oil reserves come about? What brought about the prospect of shale oil development despite all the historical polemics about snake oil peddlers, unscrupulous oil companies and land boom speculators? The start of the 1970's drive to develop Colorado's vast oil shale reserves probably can be traced back to about when I joined the prosperous oil business, after earning an engineering degree from Montana School of Mines. As noted in Chapter 1, about 1960 the Shell Oil geologist M. King Hubbert predicted that US oil production would peak within about 10 years after 100 years of contributing to the making of this country as the strongest in the world.[iii] Up until then, the industry had been grandly and gloriously finding and developing domestic fields, transporting, refining and marketing all the oil needed for fighting world wars and "military actions", at the same time building our economy into one envied by all the world. It seemed unbelievable that this idyllic situation could possibly end. As a result, Mr. Hubbert's predictions generally fell on deaf ears. And very little concern was exhibited by anyone because oil continued to be found and marketed and prosperity reigned. Automobiles for example were exemplified by powerful, spacious, luxury cars or "muscle" cars. Gasoline mileage was not of concern to any driver or purchaser. Speed, acceleration rate and roominess were what one looked for when shopping for transportation in the 1960's. It was not until considerably later that real concern for oil availability began. US oil production did not begin to decline until about 1970 (exactly as predicted by Hubbert), while at the same time oil consumption continued to rise. (Remember the emphasis on automobile power and size, rather than efficiency, which prevailed.) To make up the difference, oil imports to the US from around the world increased dramatically. By 1973, we were importing nearly 40 % of our petroleum. It might be interesting to put this in some sort of perspective. In 1973, the world was using about 57 million barrels of oil each and every day of the year. Of that we USers were consuming about 17 million barrels of oil per day, importing about 6.5 million barrels of that from other countries daily. At the same time our population was about a quarter of a billion souls and the world population was about 4 billion. So about 6 per cent of the world's people were using about 30 per cent of the world's oil.

It has invariably been difficult for thoughtful people to decide whom to blame for this situation. International politics have always played a significant role. Middle East and South American oil producing countries were continually pressuring the US and Europe into using more and more of their oil to obviously improve their local economies. There was a constant tug of war between those countries and the US Congress whose goal was to keep lower priced foreign oil out of the US and thus protect the local oil industry. Oil import quotas were set in the 50's, which maintained this status quo. It seems ironic that this protectionism has resulted in our growing reliance on foreign oil because of the lack of available domestic supplies which occurred at least partly because Congress kept the price of oil low.

The US style protectionist policies did not exist in Europe, simply because there were no known oil reserves to protect in non-communist Europe. In those countries, oil consumption grew even faster than in

the United States. As quoted in Yergin[iv], while oil consumption tripled
in the United States during the period of 1948 to 1972 from about 6
to16 million barrels per day, it increased in Europe from about 1 million
to 14 million barrels per day and in Japan from about 32 thousand to
over 4 *million* barrels per day. One reason for these dramatic increases
was the growing popularity of the automobile. US motor vehicle
ownership increased from 45 million to 119 million (2.6 times) over
this time frame while motor vehicle ownership in the rest of the world
increased from about 19 million to 161 million (8.5 times). Other
major contributors to the unparalleled growth in oil consumption
included the fact that oil became cheaper, due in large part to the
international economical politicking mentioned above while at the same
time coal became more expensive, due in part to militant labor
demands. Growing concern over air quality also entered into any decision
that might supplant coal with oil.

In mid-October, 1973, as part of the complex "Yom Kippur" war
between Egypt and Israel, an embargo against nations friendly to Israel
was declared by the Organization of Petroleum Exporting Countries
(OPEC) which promised to cut oil production levels monthly until
political objectives were met. As this was being announced, the US
coincidentally announced a direct aid package to Israel military forces,
which in turn precipitated a termination of all oil shipments to the US
and to the Netherlands. So we found ourselves in an unprecedented
situation with not enough oil to maintain normality and no non-OPEC
resource available to solve the problem. Gasoline shortages resulted in
people waiting in long lines for fuel and in the process losing confidence
in their future ability to maintain the quality of life to which they had
become accustomed. It didn't help, in my opinion, that Watergate
maneuvers filled the newspapers and laid another layer of doubt on our
future.

The President and the government did what it could, given the
situation. Talk and action about energy conservation became prevalent.
An Energy Research and Development agency was established to develop
some sort of national energy policy and strategy to mitigate this most
uncomfortable situation in which we found ourselves. A 'Project

Independence' was proposed following the model of the successful 1960's campaign to put a man on the moon.

This greatly increased sensitivity to a shortage of available energy was significantly emphasized by the price of available oil. A primary objective of OPEC was to increase the price of crude oil worldwide. As part of their Yom Kippur war maneuvers, they gained control of this price, due to their new position as key producers of a limited resource. It should be acknowledged that this situation had not been possible previously because the United States always had oil reserves available to take up the slack. These reserves were no longer available, depleted by geometric growth of oil use and curtailed to some extent by new environmental protection measures which discouraged much new exploration and development. The price of crude rose from less than $2.00 per barrel in 1970 to about $12.00 in December 1973. This increase was of course reflected in gasoline, heating oil, petrochemicals, literally everything we were using. A major period of inflation began in the US and the rest of the world. Negotiations resolved the Arab/Israel conflict, at least temporarily, and OPEC lifted the oil embargo, but, the price did not go down.

Thus, the stage was set for renewed interest in developing the US' vast deposits of shale oil. No one wanted to go through another gasoline shortage, with all its personal discomforts. Also, no one now doubted the crisis as prophesized by Hubbert nearly 15 years previously that US oil resources were limited. National security was shown to be vulnerable and the national morale was severely damaged. Could someone propose and find a solution to these dilemmas?

# CHAPTER 3

# The Reactions

THE OIL COMPANIES responded in several ways. One way was strictly defensive, for they were accused of many inter-company conspiratorial schemes. Because of the history of close cooperation between the oil companies and various levels of government, these accusations always have appeared unfair to me. However, I readily admit to a bias developed over many years of prosperous employment by these companies.

Other, perhaps more positive ways that oil companies reacted to the by-then-labeled "Energy Crisis", included buying into coal companies and beginning development of coal to oil or coal to gas projects. Shale oil plans, which had been idling for some time, were revived and given new life. One such plan was the Colony Development project near Parachute Creek. Several companies had participated in this innovative project which was based on TOSCO (The Oil Shale Company) technology. Other projects, some of which had been under some level of development since World War II and which now increased their levels of activity included Paraho (a consortium of several oil companies), Union Oil, and Occidental Oil.

Shell Oil Company chose to buy into Colony Development in 1974 and this was when I became involved. The Colony group in 1974 consisted of ARCO as operator, TOSCO, Ashland Oil and Shell. A joint venture such as this required tact and diplomacy, both between partners and in developing a relationship acceptable to the anti-trust watchdogs of the federal government. Shell was a staid, conservative company and was probably overly conservative with their various interfaces. It was an interesting and educational experience for me.

ARCO had hired C.F. Braun, an Alhambra, California major contractor to prepare the design for the Colony project. (Interestingly enough, Braun later became a part of a Kuwait company, which may or may not have a concern about developing alternative energy supplies in the US.) Representatives of the partners would congregate in Braun's offices to review process and mechanical design offerings. It was an exciting time. Everyone, with the possible exception of some of the older TOSCO folks was learning a new business. Interestingly, it was a business that went beyond the oil and petrochemical processing, in which we were all experienced, and started introducing such things as rock crushing and revegetation of spent shale rock to our vocabulary.

Unfortunately, it didn't last long. In October 1974, the combined management of the venture companies decided to suspend operations, citing double-digit inflation, tight money supplies and lack of a national energy policy. (This scenario, in essentially identical form, will come up again and again in this discourse as we proceed.) So, we generally went back to our other assignments in our respective company offices.

However, there was a parallel effort in progress, which soon got most of us involved again. As part of the federal government's expression of concern for the energy crisis, a program was initiated to develop oil shale on federally owned lands in Colorado, Utah, and Wyoming. One of the complicating factors discussed in the books cited in Chapter 1 is that much of the known oil shale deposits are located under federal lands managed mostly by the Bureau of Land Management (commonly accepted statistics relate that about 70% of the western oil shale lands with about 80% of the shale oil resource are owned by the federal government). Often, the early oil shale development efforts consisted in a large part of trying to get the government to release some of the resources. Because of history with similar government owned energy resources such as the Teapot Dome government oil reserves in Wyoming and histrionics between private oil shale developers and the government in the 1920's, there was considerable concern about unlimited release of federal reserves for private development. Fortunately, modern day oil and coal leases had produced a way for resources to be developed within acceptable monetary and land use guidelines. However, in the 1970's, a new force was demonstrating considerable power, that is the environmental movement. These people did not want development of the Colorado plateau without comprehensive management of the general ecology. Environmental impact would have to minimized and completely understood before development began.

Notwithstanding all the above, the Department of Interior had announced a Prototype Oil Shale Leasing Program in November, 1973 as part of the federal administration's efforts to mitigate the current energy crisis. This was a carefully crafted document and agreed upon by essentially none. However, it had strong redeeming features, which kept it afloat despite continual criticism. One of these was the "Prototype" featured prominently in the title. A primary goal would be to provide information on the pioneer industry, in terms of financial viability, compliance with environmental regulations and acceptance by impacted communities. It was argued that answers to troublesome questions could only be provided by construction and operation of real plants.

A summary description of the program was provided in the February, 1976 C-b Detailed Development Plan (the terms C-b and Detailed Development Plan will be explained soon) as follows:

> "During the 50 years following the inclusion of oil shale in the Federal Mineral Leasing Act of 1920, no federal oil shale lands were leased to the public. In early 1971, the states of Colorado, Wyoming and Utah submitted reports to the Secretary of the Interior on expected economic and environmental effects of the

development of oil shale in those states. Later the same year, the Secretary of the Interior announced plans for a proposed Federal Prototype Oil Shale Leasing Program. From mid-1971 through 1973, a number of activities preliminary to a decision to proceed with the proposed program took place. Those activities included: an exploratory core drilling program; nomination of proposed lease tracts by the public; establishment of a federal oil shale task force comprised of representatives from federal, state and local governments; and the preparation and publication of a draft and final environmental impact statement by the Department of the Interior in accordance with the provisions of the National Environmental Policy Act. Following the publication of the final environmental statement for the Prototype Oil Shale Leasing Program in August 1973, the Department of the Interior published its intent to offer for lease two oil shale tracts in each of the states of Colorado, Utah and Wyoming. Each tract comprised no more than 5120 acres.

As stated in the final environmental statement, the goal of the leasing program was to:

'. . . provide a new source of energy for the nation by stimulating the timely development of commercial oil shale technology by private enterprise, and to do so in a manner that will assure the minimum possible impact on the present environment while providing for future restoration of the immediate and surrounding area.'

The Leasing Program also established the following objectives:

—provide a new source of energy that will increase the range of energy options available to the nation by stimulating the timely development of commercial oil shale technology by private industry;

—insure the environmental integrity of the affected areas, and concurrently define, describe, and develop a full range of environmental safeguards and restoration techniques that can be reasonably incorporated into the planning for a possible mature oil shale industry in the future;

—permit an equitable return to all parties in the development of this public resource; and

—develop management expertise in the leasing and supervision of oil shale resource development in order to provide the basis for future administrative procedures.

Commencing with competitive bid sales in January 1974, the Department of the Interior offered for lease the six selected tracts in Colorado, Utah and Wyoming, and during the following six months leased four of these tracts, two each in Colorado and Utah. Neither of the two Wyoming tracts received acceptable bids."

Competitive cash bidding similar to that used for oil and coal leasing was the model required by the program. The bids were for 20-year leases, renegotiable at 20-year intervals. Cash bids (called Bonuses) were payable in five equal annual installments. A bonus credit provision was included which allowed off set (non-payment of the installments to the government) of the fourth and fifth installments if appropriate expenditures were made toward development of the tract(s). Royalties were to be assessed at basically 12 cents a ton of oil shale rock processed or sold to be processed. Several conditions were attached to the royalty including both credits against early production costs and pluses or minuses according to the amount of oil per ton.

The bids were considerably higher than most people expected. Likely because of the positive tone set in the bid package as described above, bonuses were bid which ranged from $45 million for one Utah tract to over $210 million for the Colorado tract C-a. (Tracts were given the uninspired names of C-a for the first one in Colorado, C-b for the second, etc.) The tract in which I had personal interest, Tract C-b, had a high bid of $117,778,000 submitted by ARCO, Ashland

Oil, Shell Oil, and TOSCO, the earlier Colony group. Similar to the Colony operation, ARCO began as the C-b managing partner.

Continuing on with the C-b Detailed Development Plan (and paraphrasing as necessary to represent the appropriate tense) as follows:

> "The Federal Prototype Oil Shale Leasing Program was to be administered by the Area Oil Shale Supervisor (AOSS) of the Conservation Division, U. S. Geological Survey, Department of the Interior. The AOSS would have the ultimate authority for the approval of development plans. However, he would receive advice from other interested governmental agencies and the Oil Shale Environmental Advisory Panel (OSEAP). The AOSS would submit all significant proposals, including Exploration and Development Plans, to the OSEAP for comment on environmental aspects on a timely basis. The AOSS was required by the Lease provisions to react to the Lessee's approval requests without undue delay."

Numerous specific environmental controls were included in the Lease and the Environmental Stipulations. These included the following provisions:

—The Lessee shall conduct all operations . . . in compliance with all applicable federal, state, and local water pollution control, water quality, air pollution control, air quality, noise control, and land reclamation statutes, regulations and standards.

—The Lessee shall avoid, or, where avoidance is impracticable, minimize and, where practicable, repair damage to the environment, including the land, the water and air.

—A two-year baseline data collection program must be conducted under the Lease; data to be collected are to cover the areas of surface water, ground water, air quality, soils, flora and fauna.

—Environmental monitoring must be conducted before, during, and subsequent to development operations, in order to provide: 1) a record of changes from conditions existing prior to development operations as established by the collection of baseline data; 2) a continuing check on compliance with the provisions of the Lease (including the Stipulations) and all applicable federal, state and local environmental protection and pollution control requirements; 3) timely notice of detrimental effects and conditions requiring correction; and 4) a factual basis for revision or amendment of the Stipulations.

—The Lessee is required to submit, for approval by the AOSS, a Fish and Wildlife Management Plan for the Tract. The plan is to include the steps which the Lessee shall take to: '1) avoid or, where avoidance is impracticable, minimize damage to fish and wildlife habitat, including water supplies; 2) restore such habitat in the event it is unavoidably destroyed or damaged; 3) provide alternate habitats; and 4) provide controlled access to the public for the enjoyment of the wildlife resources on such lands as may be mutually agreed upon.'

—The Lessee shall . . . conduct a thorough and professional investigation of any portion of the leased lands to be used, including, but not limited to, those areas to be used for mining, processing, or disposal operations or roads, for objects of historic or scientific interest . . . .

—The Stipulations call for Lessee to control erosion, and to rehabilitate and revegetate disturbed lands. An initial Management Plan for achieving this objective is required to be submitted not less than 60 days prior to the start of mining and site preparation and is to be updated each year thereafter.

—the Stipulations specify scenic standards to be followed ' . . . in all designing, clearing, earthmoving and construction activities.'

> The Lease further requires that aesthetic values be considered
> ' . . . in all planning, construction, reclamation and mining
> operations.' In addition, all operations are to be performed so as
> 'to minimize visual impact, make use of the natural topography,
> and to achieve harmony with the landscape.'"

The Partnership had submitted, in April and May of 1974, a Preliminary Development Plan and an Exploration Plan describing the exploration and environmental baseline programs proposed to be carried out on the Tract. Those plans were subsequently approved by the AOSS and the collection of environmental baseline data began in early summer 1974, with the commencement of some programs delayed until fall.

ARCO chose to resign the management role in favor of Shell in early 1975. I was fortunate to be chosen project manager by my Shell management and my family and I moved to Denver in the spring of 1975.

# CHAPTER 4

# The C-b Shale Oil Venture

SO THERE WE were in Denver with a charge to advance a major project utilizing unproven technology over an uncharted route with many as yet undefined pitfalls. How did I react to it all? As quoted in the October 1975 *Shale Country,* a publication "provided as a public service by the leaders of the oil shale industry", I was characterized as a super optimist. I was " . . . unreal because I get so excited about this assignment. But I've never had a project fit me so well before. . . . I like the project because I feel the oil-shale business is certainly an integral part of the country's solution to the energy crisis."

In general, the above personal description fit the majority of folks who worked on the various projects attempting to develop an oil shale industry. We were excited by the prospect of being pioneers much like many of our grandparents had been. We wanted to be adventurers finding the pot of gold at the end of the rainbow. The opportunity to be on the leading edge of technology and business breakthroughs was extremely challenging. We were an eclectic mix; different nationalities, different backgrounds and different ages. But it was a productive group. These energizing conditions lasted for months, if not years, until the continual change of management and targets became demoralizing, if not laughable.

Back to the project development, we had inherited a commitment to produce a Detailed Development Plan (DDP) by November 1, 1975, one year after initiation of the collection of baseline environmental data. One of our first challenges was by a contingent of the partnership, which felt that too many concessions were being made to the Area Oil Shale Supervisor. I felt strongly about this matter and brought it immediately to a head by asking for an audience at the first available Venture Management Committee. My boss wasn't too happy about that, as he didn't really want to start his reign as Committee chairman with a controversy. He agreed, however and we had our session, the result of which was a go-ahead to continue to negotiate with the AOSS in a friendly and cooperative manner. The AOSS, Pete Rutledge, was a young (my age, which makes him young) mining engineer with impressive government experience. He and I had similar interests and backgrounds. As noted above, the AOSS was given considerably power over our progress. He also was given a very difficult charge to seek and gain approval of the Oil Shale Environmental Advisory Panel (OSEAP) on all the lessee's plans. I strongly felt that we would either try to make Mr. Rutledge's life as easy as possible or do all our project work in court rather than in engineering and construction. This initial decision by management set the operating scene and philosophy for the next 10 years of working with the government. It also helped to blunt the environmental sword.

It is appropriate to look at the OSEAP. It was a group without parallel, a new experiment in Department of the Interior advisory groups. It was established in February 1974 by the Secretary of the Interior. Membership consisted of one member each from nine bureaus and offices of the Department of the Interior: Fish and Wildlife, Geological Survey, Land Management, Mines, Outdoor Recreation, Park Service, Reclamation, and Solicitor. Other federal members included the Departments of Agriculture, Commerce, Health, Education and Welfare, Housing and Urban Development, Transportation, and the

Environmental Protection Agency. Also included were two state employees from each of the three states involved; a representative of each of the three counties involved; a person active in environmental matters from each of the states, and three additional environmental activists appointed by the Secretary of the Interior.

Thus 29 people with diverse backgrounds and interests were being asked to advise on environmental matters which might arise during the oil shale development process. It looked like a recipe for disaster, a bureaucratic nightmare which would spend all its time asking for studies and/or providing a pulpit for show and tell presentations by egotists interested only in advancing their own private agendas. By some miracle, the Secretary of the Interior chose a superman, Bill Rogers, from the department offices in Denver as executive director of the panel. Mr. Rogers established an atmosphere in the Panel, which relied on "consensus" of all parties. He would not allow an issue to be decided by a vote, but would insist on continued discussion until consensus was agreed upon. To my mind, it was one of the greatest acts of productively managing a group I've ever seen. It should be repeated that the original leasing program was very controversial, and had not gained unified support from any group. Many of the outspoken critics of the leasing program found themselves on the OSEAP and probably expected to continue their efforts to halt the development one way or another. Amazingly, Mr. Rogers charted a path through all these obstacles and did not let the OSEAP stop the program.

Because of several reasons none of which involved OSEAP, we didn't make the November 1 Detailed Development Plan deadline. Several things happened during this period, which dramatically changed the situation. ARCO and TOSCO completely stopped their participation in C-b. Cost estimates prepared by the venture team and its engineering contractors seemed to be escalating without much hope for reaching a limit. The capital cost estimate prepared for the DDP was for $921 million (1975 dollars) for a 66,000 tons per day operation. Government signals were mixed as far as what would be supported politically and economically. Partnership papers claimed that withdrawal by ARCO and TOSCO came shortly after a congressional defeat of a Federal energy loan guarantee program.

Ashland and Shell, however, opted to continue to move ahead, but not without significant reservations. The DDP was issued in February, 1976 at a cost of about $12 million, which didn't include the cost of the second year of developing the environmental baseline database. Because of the reasons stated in the above paragraph, primarily associated with cost escalation and uncertainty about government intentions, Ashland and Shell immediately asked for a suspension of the lease. The stated reason was that cores removed from C-b as part of the resource exploration indicated that rock strength was less than earlier assumed. As a result, underground pillars would have to be larger than originally thought and percent recovery would be diminished. Economic calculations, which went into establishing the tract bid were thus in question and potential profits, were endangered. The other three lease operators also began discussions with the AOSS about possible suspensions for similar reasons. The Department of Interior was willing and open to discuss these matters, possibly because they wished for the program to succeed as much as we did. In August, 1976, the Secretary of Interior announced a one-year suspension of the two Colorado tracts based on established "conservation of resource" rules. The resources noted included both mineral and environmental resources. One concern recognized at that time was that ambient air quality in western Colorado already exceeded Federal clean air standards, depending on which way the wind was blowing. Clearly, it would be difficult to build and operate oil shale plants that didn't exceed restrictions when the limits were exceeded even before any construction began.

Despite our success in obtaining a lease suspension, Shell Oil decided two months later to withdraw from the C-b project because of "the huge initial project investment required, unfavorable economics, and political—environmental uncertainties". Capital costs had escalated by about 10% in less than a year's time due to extraordinary inflation being experienced in the country (at least partially due to the significant increase in energy prices which started with the 1973 OPEC embargo) with direct impacts on equipment and labor costs. Political— environmental uncertainties cited included recently enacted oil price control legislation, the above-noted failure to pass a federal loan guarantee program, threats by the federal government to split up oil companies, proposed Clean Air Act amendments plus visible opposition from the Governor of Colorado. Economic concerns included a real doubt about the project's ability to offset the fourth and fifth bonus payments because of wording contained in the suspension document leaving great uncertainty about offsetting costs of environmental baseline work still going on during the suspension period. The offsetting of these costs made a significant impact on project feasibility economic calculations.

This left Ashland Oil as the sole remaining C-b lessee. There had been much discussion during 1976 about incorporating in-situ retorting into the project as a way of improving the oil recovery rates apparently diminished by the poor rock strength discoveries. I participated in a Shell study of a combined modified in-situ (MIS)/(AGR) above ground retort which showed improved profit. This approach had an added risk, however, of doubling the number of unproven retort technologies needed for successful plant operations. Things had progressed too far within Shell for them to change direction and the withdrawal from C-b took place. Shell closed our Denver office and moved the staff to other locations. I was reassigned back to Houston with a job on Shell's Saudi Arabian petrochemical project. My extremely helpful family picked up and moved back to Houston in time for the new school year.

A new party entered the project at this time (Fall, 1976). When Ashland realized that Shell intended to withdraw from C-b, they began discussions with Occidental Oil Shale (Oxy), who had property near the town of Debeque, Colorado called the Logan Wash site where they were developing a modified in situ process. Oxy claimed significant success on what could be considered nearly full-scale sized underground retorts at the Debeque site. The two companies entered into an agreement to attempt the joint development of C-b utilizing the MIS process. Ashland granted Oxy a 50% interest in the C-b lease in exchange for the right to use the Oxy modified in-situ process to develop the resources at C-b.

In November 1976, the Oil Shale Supervisor issued a conditional approval of the DDP submitted in February. Before work could continue, however, resolution of issues, which resulted in the one-year suspension, would be required.

It was at this time that I became re-involved with the project. Oxy contacted me through a employment consultant. After significant soul searching, anxiety, and weighing of various alternatives, my family and I decided to give up our life-long career with Shell and move back to Colorado. It gives me some pain even now to write these words, it was such a significant decision. Our middle son, who would suffer the most school based trauma going into his senior year in high school was most supportive and enthusiastic about moving, which made a real difference to me. One must realize that the alternative we were considering was an assignment in Saudi Arabia, with our children in boarding school in Europe or elsewhere. This helped to swing the vote to Colorado. My wife and two of our children stayed in Houston while I went to Colorado. The logical place for our project office seemed to be Grand Junction, on the western slope of the Rockies and closer than Denver to the oil shale deposits and our 5000-acre tract C-b. So that is where I headed in December 1976. One of my very best friends, Jim Brown, also decided to leave Shell and join the Oxy C-b team. That was a great thing for the project and me, as he brought a sense of business intelligence and common sense necessary for acceptance of the fledgling oil shale industry by the financial community. A further good thing was that my oldest son decided to earn some money between college terms and came to Colorado. So the three of us began batching in an apartment in Grand Junction. Jim's family and the rest of mine came to Grand Junction after the end of the 1976/77 school year.

Ashland and Oxy submitted Modifications to the DDP in February 1977 to the AOSS with the express purpose of answering the suspension questions. The Modification focused on Oxy's MIS experience and the effective utilization of this experience and their process in developing C-b so that resource recovery would be enhanced irrespective of rock strength problems. The preliminary capital cost for the modified in situ based plan represented in the Modification was about half that of the surface retorting option proposed in the initial DDP.

This Modification was approved by the AOSS in August 1977 just in time to start work at the expiration of the lease suspension.

# CHAPTER 5

# Industry Status

T HERE WAS LOTS of activity underway from 1977 to about 1981. Too much of it was in bureaucratic offices in Denver, Salt Lake City and Washington, D.C. However, significant work was being performed in energy company and contractor offices around the country, as well as on the various sites in Colorado and Utah. To keep things in perspective, this chapter will present a review of first, the C-b project's status at this time and second, a briefer discussion of the rest of the other recognized projects being worked on.

## The C-b Project

The C-b project developments to date and the basis for the project at this point in time were described in the approved DDP modification, *to wit*:

*"Summary of Lessee's Commercial Program*

1.  Summary of Activities:

> The Tract C-b development program will extend from
> September 1, 1977, starting with site preparation, to
> September 1982, when the surface facilities and mine will
> be capable of full-scale production.

> The property will be mined by developing successive areas
> called panels. A panel will contain several groups of retorts.
> These groups are referred to as clusters or retort clusters. A
> retort, the basic unit, is a column of rubbled shale
> containing a certain void fraction bounded by undisturbed
> pillars. It is envisioned that a group or cluster of eight
> retorts will be processed concurrently. Clusters will be
> brought on-line in such a sequence that the number of
> clusters operating at one time within a panel will produce
> the required quantity of shale oil.

a.  Site Preparation and Shaft Sinking

> The first site activities starting in September 1977
> will be to get shaft sinking underway. Underground
> development can proceed only after shafts are
> available.

> While the shaft-sinking operations are being mobilized,
> certain site preparation and preconstruction activities
> will proceed. These will include preparation or extension
> of service roads, construction of water storage to receive
> underground water from initial dewatering operations,
> and necessary grading for temporary construction
> facilities, fencing, etc.

The Lessee will construct five vertical mine shafts on the tract: the ventilation/escape shaft (12 feet in diameter), the temporary gas shaft (6 feet in diameter), the production shaft (34 feet in. diameter), the service shaft (34 feet in diameter), and the gas shaft (34 feet in diameter).

Of the four major shafts, the first to be completed will be a 12-foot-diameter shaft 500 feet from the northern property boundary. This shaft will perform a dual service during the development period. First, it will serve as a temporary production shaft until the main 34-foot-diameter production shaft becomes available, and later it will serve as a ventilation/escape shaft for the mine.

b.  Ancillary Development

As noted above, the 12-foot-diameter shaft will be completed first. While the 34-foot-diameter shafts are still being constructed and major mine headings and levels are being opened up, an ancillary facility will be constructed in the vicinity of the 12-foot-diameter shaft. This facility will include two or more commercial-sized retorts, retort clusters and attendant equipment to operate them. This will permit the processing of a thick section of high-grade oil shale, establish the environmental monitoring procedures, obtain operating experience, and provide a site for the training of mine and processing personnel.

Later, as the main shafts and other full-scale plant units become available, this ancillary facility will be incorporated into the full-scale facility.

c. Full-Scale Modified In Situ Plant and Operations

The construction period of the full-scale commercial facility will begin with shaft sinking in September 1977 and end September 1982 when the first cluster within the commercial panel is kindled. From this time on mining will no longer be a preproduction activity as part of the construction program; it will then become a production operation. As more clusters are brought on-line production will increase and full-capacity production (57,000 bbl/day) will be achieved in about 12 months, or by September 1983.

Within the 5-year construction period, the ancillary retort facilities will be developed and processed as briefly described above. The ancillary facilities will use the shafts and some increments of the full-scale oil/gas processing units for their operations.

After key development drifts are driven at all levels around the panel, the principal work of retort preparation will begin in May 1981 and continue to September 1982. The construction of the major oil/gas processing units and general facilities will occur between January 1981 and September 1982.

Operations will continue from September 1982 to about the year 2040, when the resource interval will be exhausted.

(1) Description of Mining

The mine will be developed for commercial operation in two basic degrees. The first will include sinking two small shafts of approximately 6 and 12 feet in diameter to prepare an ancillary development of commercial size in situ retorts for operation. This will permit establishing environmental monitoring procedures, obtaining operating experience and providing a site for training of operating personnel. The second will include sinking of three 34-foot-diameter (production, service, and gas) shafts and developing the full-scale mine for commercial operation of in situ retorting.

Multiple levels of access are required for full-scale in situ retort development and operation. The uppermost provides process access and acts as an air plenum for distribution of air supplied to the retorts. Access levels penetrate the retorts below the air level to allow creation of voids required for rubbling. The production level is below the retort and is used for withdrawal of material removed during development and creation of retort voids. It also allows gathering of the oil and water produced by the retorts in operation.

Immediately below the production level is the gas level, which contains the gas collecting drift through which the gaseous products of combustion are collected and carried to the gas shaft and hence to the surface facilities.

The initial small shafts will be retained and used for ventilation, escape ways and mine dewatering.

The large-diameter production and service shafts will be used for ventilation and to supply air for the retorting process.

(2) Description of In Situ Retorting

The Modified In Situ Process is a method of retorting oil shale underground as compared to conventional surface retorting in which shale is mined, transported to a centrally located surface plant for retorting, and the spent shale ultimately disposed.

In the Modified In Situ Process, retorts are created by mining out only enough shale to provide a void fraction for rubbling the remaining shale by blasting and to provide permeability for gas flow during operation. These in situ retorts consist of groups or "clusters" of eight 200-ft x 200-ft x 310-ft-high rubble columns or chimneys. Undisturbed pillars function as control partitions between operating retorts.

The processing of a cluster of retorts consists of several steps. First, a retort within a cluster is kindled from the top by externally fueled burners. When the temperature at the top of the retort is sufficient to sustain reaction, the burners are shut off and a regulated mixture of air and steam is drawn into and through the retort by exhaust blowers on the surface. Residual organic material is combusted with the air in the feed gas. The hot combustion gases flow down through the retort and supply heat to the raw unretorted shale below. As the shale is heated, the organic material or kerogen decomposes into oil vapor and other gases that are carried along with the combustion gases while some residual organic material remains in the rubble. Steam in the feed gas acts as a diluent to the oxygen in the air to control the reaction temperature and reacts with some of the residual organic material, forming carbon monoxide and hydrogen to improve the heating value of the product gas. Some of the mineral carbonates in the shale are also decomposed to carbon dioxide gas and mineral oxides. As the gas mixture flows down through the retort, it preheats the raw shale. At the same time, the oil and some of the water vapor are condensed. Product liquids and gas leave the bottom of the retort and move to the surface for further processing as product oil, produced fuel gas, and water.

As retorting progresses, the combustion and retorting zones move slowly down through the in situ retort. Between 7 and 8 months are required to process a cluster. When retorting is complete, the air and steam feed are stopped and the in situ retort is closed off. The spent shale remains underground with no need for surface disposal.

Surface process facilities consist only of oil-water separation equipment, exhaust blowers, a sulfur removal unit for treatment of product gas, and boilers to produce process steam from fuel values in the product gas.

(3)   Water Supply and Disposal

Commercial operation will require about 2500 gpm of water. Present information indicates that water supply and requirements will essentially be in balance. Any deficiency will be made up either from on-tract wells or from off-tract sources.

If a surplus of water occurs, the excess will be stored and then either reinjected or treated and released.

(4)  Off-Tract Facilities

A staging area and terminal facility will still be required for off-tract facilities. The access-road corridor as outlined in the original DDP still applies.

With the expected arrival of Alaskan crude on the West Coast, it is generally thought that the Four Corners area and the Texas-New Mexico crude oil lines will soon be filled with Alaskan crude. The proposed La Sal pipeline described in the DDP would deliver shale oil into this area and on to a southern or a western market.

This DDP Modification presents an additional alternate pipeline route which will go generally north and east to tie in with an existing pipeline system which is reported to have 70,000 to 80,000 bbl/day of available space in Casper or Guernsey, Wyoming. This will allow shale oil to be delivered into the crude oil deficient northern tier and central states.

d.  Postoperations

Decommissioning of the plant and rehabilitating of the site are discussed in the original DDP, and apply in general to this mining and retorting scheme.

2.  Resource Recovery

The shale interval from which oil will be extracted (310 feet) contains about 3.0 billion bbl in place. Of this, 1.2 billion bbls will be recovered by the In Situ Process. The productive life has been calculated at approximately 56 years. This is based on a daily throughput rate of approximately 57,000 bbl/day from the modified in situ facility. If the option is exercised to surface retort the mined out shale, the total recovery would be about 1.65 billion barrels.

Nahcolite and dawsonite are present to a very minor extent in the shale zones, but these latter deposits are so dispersed and represent such low concentrations that it is deemed uneconomic to recover them.

3.   Schedule, Manpower and Cost Estimates

a.   Schedule

In the last quarter of 1977, all activities necessary for major shaft sinking, site development, and final detailed engineering design will commence. The development sequence spans from the start of engineering and procurement (2/1/77) to beyond the starting date of the full-scale commercial in situ facility (9/1/82).

Milestone dates are as follows:
9/1/77 Start Construction Activities
5/1/80 Start Initial Retorts of Ancillary Facility
9/1/82 Start Operation of Full-Scale In Situ Plant

b.   Manpower

Over a 6-year period from 1977 to 1983 there will be a gradual increase in permanent site personnel to 1,600. After 1983 this number will remain constant. In addition to permanent personnel there are two periods during which temporary construction personnel will be onsite. The ancillary development stage is predicted to have a manpower peak of 1,180 people. The In-situ construction stage manpower peak is estimated to be 2,900. In each period the buildup to peak personnel occurs over a period of about 2 years and the drop off in about 1 year.

c. Cost Estimates

Definitive cost estimates for a commercial oil shale project can be made only after further experience has been gained and more information is gathered on site characteristics. Any estimates made at this point are based upon preliminary engineering design and project scheduling. Order-of-magnitude estimates have been made to provide a general projection of the capital and operating costs, which are expected to occur during the life of the project. These estimates assume favorable economic and political climates, which allow proceeding on the planned schedules. Individual components of costs may vary significantly or may be eliminated in actual practice.

Capital cost estimates prepared for this document show overall project costs through the time of plant construction, but exclude interest during construction and deferred capital expenditures. Project estimates do not include escalation costs. Order-of-magnitude capital estimates for the modified in situ system are shown in Table I-A.

### Mine

| | |
|---|---|
| Shafts, Hoists & Appurtnances | $38,053,000 |
| Mobilization | $9,537,000 |
| Equipment & Spares | $21,450,000 |
| Conveyors & Materials Handling | $19,003,000 |
| Labor | $18,360,000 |
| Other Materials | $29,328,000 |
| Taxes | $965,000 |
| Subtotal | $136,696,000 |

### Oil & Gas Processing

| | |
|---|---|
| Steam Generation Plant | $21,060,000 |
| Water Treatment Plant | $36,250,000 |
| Gas Treatment Plant & Heater-Treater | $10,200,000 |
| Other Major Equipment | $15,568,000 |
| Other Materials | $12,405,000 |
| Taxes, Spare Parts & Miscellaneous | $3,036,000 |
| Field Indirects | $4,519,000 |
| Subtotal | $103,308,000 |

### General Facilities

| | |
|---|---|
| Emergency Generating Equipment, Package Boiler, Hoist, Cranes, Air Compressors, Storage Tanks & Misc. | $9,532,000 |
| Direct Materials, Concrete, Electrical, Insulation, Roads, Other Civil | $61,088,000 |
| Other Direct Costs, Spare Parts, Taxes | $2,984,000 |
| Fields Indirects | $10,378,000 |
| Subtotal | $83,982,000 |

| | |
|---|---|
| **Total Installed Equipment** | **$323,716,000** |
| Contractors, Engineering. Home Office, Fees, Working Capital & Contingency | $118,893,000 |
| **Total Capital Investment** | **$442,609,000** |

4.  Options and Alternatives

> Several options may present economically attractive alternatives as a supplement to the basic program described in this DDP Modification. The principal options envisioned at this time are:
>
> • Surface Retorting of Mined-Out-Shale
> • On-Site Electric Power Generation from Low-Btu Product Gas
> • Alternative Water Supply and Management
>
> Each of these cases is highly dependent upon economic and technical factors that cannot be determined at this time. However, considering the length of the program, it is conceivable that many of these factors will become definable, resulting in an alternate course that will be advantageous to the program.

5.  Plans and Procedures for Environmental Protection

> The environmental protection plans described in the original DDP are generally applicable to the revised program. The introduction of in situ retorting requires additional plans for protecting subsurface waters from pollution and for preventing fires underground. Fires are prevented by the low permeability and high heat capacity of the rock, by operating the retort system at negative pressures relative to ambient conditions, and by controlling the air supply to each retort. The environmental control and monitoring procedures will be tailored to protect the environment from significant adverse changes.

## 6.  Due Diligence

> The schedule presented in this DDP Modification shows
> production commencing in 1980 from an ancillary facility.
> This is considerably earlier than outlined in the DDP and
> allows the Lessee to meet the due diligence requirement of
> Section 10(aJ (3)) of the lease.

> The time schedule for development of the lease is believed
> to be attainable assuming no unforeseen delays. The
> schedule is based on restarting shaft sinking and dewatering
> in September 1977 when the suspension of operations
> ends, subject to approval by the AOSS of the Modified
> DDP plan prior to September 1977 and securing all
> required permits. Major uncertainties still exist, and the
> potential for delays is ever present. The likelihood of
> reducing the schedule is remote and the possibility of it
> being extended is high."

It can be seen that this plan was very ambitious, based on unproven
technology and assumed "favorable economic and political climates".
Review of the other shale oil projects underway in this time period
shows that all were in the somewhat the same shape. Interestingly
enough, the C-a project which commanded the highest industry bid
and was characterized as an open pit surface mine with the oil being
extracted from the shale rock in surface retorts was now proceeding on
a modified in-situ development plan. As stated in an industry-wide
report, C-a, which was a general partnership of Gulf Oil Corporation
and Standard Oil Company (Indiana) and which had adopted the
corporate name Rio Blanco Shale Company was engaged in a three-
part development program. In addition to the MIS effort, they were
investigating the merits of various surface retorting processes which
could extract the oil from the shale rock removed to form the
underground MIS retorts. In addition, they were continuing to
investigate and spend engineering money for the open pit development

originally envisioned. Their publicly stated position was that federal land adjacent to and in addition to the 5000-acre lease was needed for process facilities and disposal of overburden and processed shale. They believed that it was understood at the time of the lease that this land would be made available at the appropriate time, but the Department of Interior had ruled that they did not have the authority to lease off-tract land. Fruitless efforts to pass legislation to remedy this situation characterized the project status and further exemplified the continuing tension between the government and oil shale developers. Rio Blanco tentatively projected a commercial production rate of 50,000 or more barrels per day by 1987 if everything went as they assumed.

To further show that energy companies were taking the oil shale business seriously, the following are excerpts from the above-mentioned industry wide status report.

## Update on Individual Shale Projects

### *Atlantic Richfield Company*

The ARCO Coal Company division of Atlantic Richfield currently holds approximately 24,000 acres net interest in oil shale. Following the sale of its 60 per cent interest in the Colony Shale Oil Project to Exxon, U.S.A., ARCO Coal initiated an environmental assessment and exploration program on these remaining interests. No specific commercialization plans have been announced.

### *Chevron Shale Oil Company*

Chevron Shale Oil Company, a wholly owned subsidiary of Standard Oil Company of California, proposes to develop its Clear Creek property to produce 100,000 barrels per day of shale oil by the mid 1990's. The property is located in Garfield County approximately 40 miles northeast of Grand Junction and 26 miles north of DeBeque.

The proposed Chevron project will include the development of a combination underground/open pit mine, adjacent surface retort and upgrading facilities; a water system built in conjunction with Getty Oil Company and Cities Service Company consisting of an intake structure in the Colorado River, water pumps and pipelines, a storage reservoir in the Roan Creek Basin and support facilities including pipelines; and roads and community facilities for the Chevron Shale Project.

Chevron has developed an oil shale retorting process called the Staged Turbulent Bed (STB) in its Chevron Research Company's laboratory in Richmond, California. The process has been tested in a one ton per day pilot plant at Richmond. Chevron Research plans to construct a 350 ton/day semi-works development plant on property adjacent to Chevron's refinery near Salt Lake City, Utah. Construction of this facility is expected to begin in late 1981. The semi-works plant will begin operation in 1983.

Chevron recently traded a 30 percent interest in the Clear Creek property for an interest in selected Conoco coal properties, giving Conoco a 30 percent interest in the proposed Chevron project.

Chevron proposes to construct the 100,000 barrels per day project in stages beginning with a 10,000 to 25,000 ton/day retort. Construction of this first stage is expected to begin in 1984 following a 6 to 12 month period of successful operation of the semi-works plant near Salt Lake. Additional retorts, upgrading and support facilities will be built in stages until the full 100,000 barrels per day plant is completed in the mid 1990's.

Chevron initiated an intensive environmental baseline study in 1979. The study will provide data necessary to plan an environmentally sound staged development. In addition, Chevron is conducting numerous extensive studies to insure the orderly development of the infrastructure required to support the proposed Chevron project.

Chevron recognizes that with the significant economic benefits provided from the development of its oil shale resources comes the responsibility to minimize the resulting environmental and socioeconomic impacts. The project is designed to accomplish this balance.

## The Colony Shale Oil Project

The Colony Shale Oil Project is a joint operation of Exxon Corporation and The Oil Shale Corporation, a subsidiary of Tosco Corporation. Exxon Company, U.S.A., a division of Exxon Corporation, serves as operator for the project. Colony's semi-works facilities and proposed commercial mine and plant are on private property located 15 miles north of Parachute, Garfield County, northwestern Colorado.

Colony has conducted extensive field tests of mining operations and of oil shale retorting at the semi-works facility located at the project site. These feasibility tests were designed to demonstrate equipment operability, confirm product yields, and provide engineering data required for a commercial design. In 1972, it appeared oil shale commercialization was nearing reality, and commercial design activities were begun. C.F. Braun & Company, in connection with Fluor-Utah, was chosen to conduct a definitive engineering study and develop a cost estimate. This $12 million engineering study required 400,000 man-hours of effort, and for the first time resulted in a definitive evaluation of the costs of producing shale oil.

Unfortunately, as the final decision approached for the start of construction of a commercial facility, equipment and construction costs skyrocketed following the 1973-1974 Arab oil embargo. The cost of the 47,000 barrel per day Colony facility, projected at $450 million in 1973, escalated to more than $800 million by mid-1974. Thus, plans for May 1975 start of field construction were suspended in late 1974 because of economic uncertainties and the risk of proceeding without an encouraging energy policy for synthetic fuel development. However, project activities, such as community planning, environmental studies, and permit work prerequisite to a construction start, were continued.

As a result of Colony's preparations for construction of a commercial facility, over $3 million was spent to conduct more than 100 separate environmental studies between 1969 and 1975. These included preparation of a 20 volume, 6,500 page Environmental Impact Analysis describing in detail the existing environment and assessing the effects of commercial operation. This information was used to assist the Bureau of Land Management in the preparation of an Environmental Impact Statement, which was issued in final form in mid-1977.

In 1980 Exxon Corporation purchased Atlantic Richfield's 60 percent interest in Colony and preparations began immediately to develop a commercial facility.

Major contracts have been awarded for engineering design and construction of mine, plant, upgrading and attendant facilities capable of producing 47,000 barrels per-day of shale oil. The cost is in excess of $2 billion dollars and scheduled start-up is late 1985 or early 1986.

Oil shale will be mined using the room-and-pillar method. This requires large columns or pillars of oil shale be left in place to prevent overlying rock from caving into mined-out rooms. Mined shale will be crushed and conveyed to a stockpile near the retorting facility. Shale from this stockpile will then be sent to a second crushing plant to reduce the size of the ore to one-half inch in diameter or smaller. The finely crushed ore will then be fed to one of six TOSCO II retorts operating in parallel.

The TOSCO II retorts are designed to use continuously circulating hot ceramic balls to recover shale oil. Balls, which have been heated to about 1200F, are mixed with crushed shale in a rotating drum. Heat from the balls converts kerogen to shale oil, which at the high temperatures involved, is in vapor form. The balls, now at about 900F, are separated from spent shale in a screening step, then reheated for reuse in processing additional shale. Hot spent shale will be cooled, moistened, and conveyed to the disposal embankment. Disposal will include spreading, compacting, and revegetating.

After the shale oil vapor leaves the TOSCO retorts, it is cooled and condensed. Colony plans to upgrade this raw shale oil on-site. This processing will include treating with hydrogen to remove nitrogen and sulfur compounds. The nitrogen is converted in the upgrading process to ammonia, a valuable by-product for fertilizer production. The products which will be produced at the Colony facility include 17.2 million barrels/year of oil projected to be used as transportation fuels, such as hydrotreated shale oil, gasoline, jet fuel, diesel fuel or petrochemical feedstock. Also, 52,000 long-tons/year of Sulfur will be produced which is expected to be marketed for rubber manufacture and production of chemicals and fertilizer. Finally, 41,000 long-tons/year of ammonia will be produced for agricultural fertilizer production.

The 47,000 barrel-per-day oil shale facility will provide permanent jobs for approximately 2,000 employees. Additional jobs will be created in the communities supporting the oil shale development. To accommodate anticipated population growth, construction has begun on Battlement Mesa, a new community to be built just south of the present town of Parachute. Battlement Mesa will be an open community which could ultimately include 7,100 dwelling units which will provide a pleasant place to live for up to 25,000 new residents. In addition to a wide range of housing, the development will include schools, churches, shopping facilities, a golf course and other entertainment and recreation facilities.

## The BX In Situ Oil Shale Project
### Equity Oil Company

Equity Oil Company is operating an in situ oil shale research project known as the BX In Situ Oil Shale Project (BX Project). This project, with a total cost of $10.7 million, is being performed under a cooperative agreement between Equity Oil Company and the U. S. Department of Energy.

The project goal is the testing and development of a process to produce oil from oil shale in situ by injecting superheated steam at design conditions of 1000F and 1500 psig into the "leached zone" of the Parachute Creek area of the Green River Formation in the center of the Piceance Creek Basin in Northwestern Colorado. The BX Project is located on a 1000-acre tract of private property owned by Equity. On this tract the "leached zone" contains an estimated 1,125,000 barrels of shale oil. It is estimated that the total "leached zone" oil shale resource of the Piceance Creek Basin may be as large as 275 billion barrels.

More specifically, the objective of this project is to demonstrate the technical feasibility of using superheated steam as a heat-carrying medium to retort in situ the oil shale in the Green River Formation "leached zone," and to provide a mechanism for the recovery of this oil with a minimum impact on the environment.

The steam is injected into the "leached zone" through an array of eight injection wells. The steam, water, oil and gas produced from the "leached zone" are recovered from five production wells.

The project has been in operation continuously since September 1979. Through February 1981, 625,000 barrels of water as steam have been injected into the eight project injection wells.

The following conclusions can be reached from operations of the Project through February 18, 1981:

1.  Superheated steam can be injected into the "leached zone" at a reasonable rate that is principally controlled by injection pressure, the number of perforations in the injection well, and the withdrawal rate by production from the "leached zone."

2.  When superheated steam is injected into the "leached zone" the steam gives up its heat to the zone and appears to move out from the injection wells in a principally horizontal direction and in a basically isotropic fashion.

3.  Formation heating in the "leached zone" by superheated steam injection has resulted in the in situ retorting of some of the oil shale in the "leached zone" and the oil shale appeared in measurable quantities in production wells.

4.  Oil produced to date by in situ retorting in the field has physical properties similar to those of oil produced in a laboratory experiment run to simulate actual field experience.

The Project is now scheduled to continue operations through January 1982. The goal for the remainder of the injection period is to maximize steam injection rate, injection temperature, fluid production rate, and oil recovery. If present trends continue, it is expected that adequate data can be collected to evaluate the technical feasibility of the process.

## Geokinetics, Inc.

Geokinetics, a publicly owned corporation, holds oil shale leases on 30,000 acres of land in Uintah County, Utah. Total in place reserves are calculated at 1.7 billion barrels.

The Company was organized in 1969. Since 1975 Geokinetics has been developing the Geokinetics Horizontal In Situ Retorting Process at its field test site 70 miles south of Vernal, Utah. In the Geokinetics process, a pattern of blast holes is drilled from the surface, through the overburden, and into the oil shale bed. The holes are loaded with explosives and fired, using a carefully planned blast system. The blast results in a fragmented mass of oil shale, with a high permeability. The void space in the fragmented zone comes from lifting the overburden, and producing a small uplift of the surface.

The fragmented zone constitutes an in situ retort. The bottom of the retort is sloped to provide drainage for the oil shale to a sump where it is lifted to the surface by a number of oil production wells. Air injection holes are drilled at one end of the retort, and off gas holes are drilled at the other end. The oil shale is ignited at the air injection wells, and air is injected to establish and maintain a burning front that occupies the full thickness of the fragmented zone. The front is moved in a horizontal direction through the fractured shale towards the off gas wells at the far end of the retort. The hot combustion gases from the burning front heat the shale ahead of the front, driving out the oil, which drains to the bottom of the retort, where it flows along the sloping bottom to the oil production wells. As the burn front moves from the air intake side of the retort to the off gas wells, it burns the residual coke in the retorted shale as fuel. The combustion gases are recovered at the off gas wells. This gas is combustible and can be used for power generation. Progress of the burn front is monitored by thermocouples set in thermocouple wells.

The current schedule is to complete the research and development program by the end of 1982. The first production unit will consist of a 2,000 barrel-per-day operation at the same location as the test site. Additional 2,000 barrel-per-day production units will be brought into production on other company leases at the rate of about one per year. Employment will be about 70 persons for each production unit. Because of the small work force, Geokinetics does not expect to have a significant impact on the local community services.

### Mahogany Shale Project

In April 1980, Phillips Petroleum Company leased from the claim owners a block of unpatented oil shale mining claims located in the Northwest part of the Piceance Creek Basin. The lease grants Phillips the right to mine and develop oil shale and associated minerals. The lessors plan to apply for patents to such claims.

Phillips is planning to develop the property for oil shale production, but commercial production will probably not be realized before the early 1990's. Specifics on plans for development are not yet determined.

### Multi Mineral Corporation

Multi Mineral Corporation (MMC), a wholly owned subsidiary of Charter Oil Company, was formed in September of 1978 for the purpose of demonstrating the integrated in situ (iiS) process and obtaining a lease on which to carry out a commercial multi-mineral shale oil project.

In May of 1979, Multi Mineral Corporation began experimental mining in the Saline zone at the U.S. Bureau of Mines' Experimental Oil Shale Facility under a cooperative agreement with the Bureau of Land Management and the Bureau of Mines. The facility, located on Horse Draw in the Piceance Creek Basin about 9.5 miles east of federal lease tract C-a was built by the Bureau of Mines in 1977-78. It consists of a lined, 8-foot diameter shaft, 2,350 feet deep, and related facilities.

This shaft is the only access to the deep oil shale beds of the Basin which are known as the Saline zone. The experimental mining consisted of two separate activities: 1) mining four test rooms around a pillar in a bedded nahcolite zone, and 2) testing of the stope or retort rubblization portion of the integrated in situ process. The four test rooms in the bedded nahcolite horizon have been mined and instrumentation has been installed to monitor rock stability. The second activity, stope rubblization, is set to be completed by June 1981.

A second program for a demonstration of the integrated in situ process at the Horse Draw facility was submitted to the Bureau of Mines in December of 1979. The Bureau determined that approval of such a program would require an Environmental Impact Statement. A contract was issued and the statement had reached the draft stage when the Mine Safety and Health Administration determined that a second shaft would be required prior to any underground testing of the retorting and leaching portions of the process. This requirement for a second shaft effectively rendered the program unfeasible due to the extra $5 million to $10 million that would be required.

A decision was thus made to build a research facility in Grand Junction with a retort of sufficient size to allow a test program using shale of the same particle size anticipated for the commercial process. This testing is necessary to fine-tune the factors involved in retorting large diameter particles through a hot gas to solids heat transfer. Almost all of the previous research and development in oil shale has been performed on Mahogany zone shale using smaller particles and either combustion or solid to solid heat transfer. The Superior Oil Company's Circular Grate Retort has been demonstrated on Saline zone shale using hot gas to solids heat transfer, but the particle size was much smaller than that proposed for the iiS process. The research facility will contain two retort vessels-a small 500 pound charge unit and a larger unit approximately 8 feet inside diameter and 40 feet process height. The larger vessel will be capable of an 80-ton batch charge. A unique design has been incorporated into the larger retort to allow the actual commercial condition to be closely simulated. It will also be possible to test the leaching portion of the iiS process in the larger retort. Initial test runs are planned for August, 1981.

Multi Mineral Corporation is involved in the current Department of the Interior effort to expand the Prototype Oil Shale Leasing Program. Over 95 percent of the resource that could be developed by the iiS process is currently under federal control while most of the non-federal land is in 40-acre homesteads, along the creek bottoms. Multi Mineral Corporation expects to be in position to carry out a modular demonstration of the iiS process as soon as a shale lease is obtained, followed by a commercial development for production of nahcolite, shale oil, soda ash, and alumina. However, the existence of sodium leases in the Basin allows development of a nahcolite-only project to proceed before an oil shale lease is issued.

In June, 1980, Multi Mineral Corporation acquired the rights to an 8,300 acre federal sodium lease from Industrial Resources, Inc. This lease, issued in 1971, is located in the north central portion of the Basin (about 8 miles east of tract C-a). Exploration drilling of the lease was begun immediately and in January 1981, a mining plan was submitted to the U. S. Geological Survey for approval. The proposed schedule for the nahcolite mine calls for shaft sinking to begin in early 1982, initial production of around 400,000 tons per year in 1984 with a production level of 1 million tons per year to be reached in 1985.

Peak employment is 440 persons with 350 persons located on site and 90 persons involved in the transport of the nahcolite product. Specific plans to address the socioeconomic aspect of this project will be developed through negotiations with the local entities involved once the timing of the development is set.

### Paraho Development Corporation

Paraho Development Corporation with its patented technology, has produced over 4.6 million gallons of crude shale oil under environmentally acceptable conditions. Paraho's initial oil shale work began in the early 1970's when a group of 17 industry sponsors (Sohio, Southern California Edison, Cleveland-Cliffs, Kerr-McGee, Gulf, Shell, Amoco, Exxon, Davy McKee, Mobil, Sun, Webb Venture, Texaco, ARCO, Phillips, Marathon, Chevron) participated in the Paraho-managed demonstration program. This demonstration program, privately funded at a cost of over $10 million, was carried out at the Anvil Points Oil Shale Mine and Retorting Facility near Rifle, Colorado. This facility is leased by a Paraho subsidiary from the Department of

Energy. The program successfully demonstrated the Paraho technology in a pilot plant and semi-works retort.

The Paraho retort is of the direct heat type—that is, both combustion and retorting occur in the same retort vessel. The feed shale enters the top of the retort, falls by gravity and is discharged at the bottom as retorted shale ash. The retorting action occurs in the top half and the combustion action in the middle of the vessel. A recycled gas combined with fresh air enters the bottom of the retort where cooling of the retorted shale takes place. No water is used. The hot gaseous products of combustion pass upward through the retort in countercurrent flow to the shale and are cooled before discharge. The gas carries the condensed shale oil in the form of vapor and mist out of the top of the retort to the separator. The combustible gases produced from kerogen decomposition are mixed with flue gases and form a low Btu gas by-product along with the oil. The low Btu gas, with about one tenth the heat content of natural gas, can be used for electrical generation but it must be used at the site because transportation costs would make it too expensive to use elsewhere.

Following the demonstration program, Paraho was selected by the United States Navy's Office of Naval Research and the Department of Energy to produce up to 100,000 barrels of shale oil. The purpose of this program was to produce and ship the largest amount of crude shale oil ever to be refined in a modern commercial refinery. This Paraho crude shale oil was refined by The Standard Oil Company of Ohio into petroleum products meeting strict military specifications.

The success of Paraho's technology on U.S. shales has given the firm an international reputation for having a successful oil shale technology. Many foreign countries are seeking Paraho's advice and expertise in the development of their oil shale deposits. Paraho has performed two pilot plant operations on Israeli shale and recently completed a pilot plant operation on Moroccan shale. Work on foreign and domestic shales is continuing at Anvil Points.

Operations at Anvil Points by Paraho have successfully produced large amounts of crude shale oil in long continuous operations with high yields and good service factors. More importantly, the operations confirmed the environmental acceptability of the Paraho process.

Paraho's operations have met all of the strict federal and state air and water emissions standards.

Environmental problems have been minimized using the Paraho technology. There are no fugitive dust problems, no expansion or "popcorn" effect problems with the retorted shale, and finally no water problems. The Paraho retort itself requires no water. On a commercial basis, approximately one barrel of water would be required for every barrel of oil produced. This water would be used in mining for dust control, reclamation for initial vegetation growth and for plant needs.

The successful completion of the demonstration program, and then the production program, led to the next logical step in oil shale development. In July of 1979, Paraho responded to a solicitation by the Department of Energy for proposals to design and demonstrate a commercial, full-size oil shale retort or module. A module consists of a mine to produce the oil shale, a single, 10,000 barrel per-day retort to recover the shale oil and gas, and all the supporting equipment required. Paraho's proposal was selected and the module contract was executed with the Department of Energy (DOE) in June 1980. This contract calls for an 18 month program costing approximately $9 million.

Paraho manages the program. Costs are being shared approximately equally by the DOE and a group of industry sponsors: ARCO Coal, Chevron, Conoco, Davy McKee, Husky, Mobil, Mono Power (Southern California Edison), Phillips, Placid Refining, Sohio, Sun, Texaco, Texas Eastern, and The Cleveland-Cliffs Iron Company. Sohio, McKee, Cliffs, and three environmental firms serve as sub-contractors to Paraho.

The DOE, in September of 1980, awarded Paraho a grant of over $3 million to prepare a commercial feasibility study covering the expansion of the single module retort into a commercial three-retort facility capable of producing over 30,000 barrels per day of shale oil. The successful completion of the module design and engineering effort and the commercial feasibility study should lead, with proper financial incentives from either the DOE or the Synthetic Fuels Corporation, to the construction of the Paraho-Ute Facility as early as mid-1982. Operations would begin in late 1984. The site of the facility is some 40 miles southeast of Vernal, Utah, on state oil shale leases and private subleases controlled by Paraho.

*Texaco, Inc.*

Economical extraction of liquid and gaseous hydrocarbons from oil shale and tar sands without mining may prove possible if a new process under development by Texaco Inc., Raytheon Company, and its subsidiary, The Badger Company, Inc., is successful.

The technique employs radio-frequency electric fields to heat deposits containing immobile heavy hydrocarbons. This process produces liquid and gaseous hydrocarbons in place, without the necessity of mining, retorting, and waste disposal. Field tests are currently being conducted on Texaco-owned oil shale property in Uintah County, Utah.

The key feature of the new process is the ability to develop heat throughout large quantities of oil shale, which is normally a very poor heat conductor. The new process, using controlled application of radio-frequency electric fields, does not depend upon the shale itself to act as a heat conductor.

This form of true in situ recovery of hydrocarbons offers advantages over the mining of oil shale deposits for surface processing. The surface land area is not disturbed and therefore requires no restoration after the hydrocarbons have been extracted. Moreover, the process eliminates major waste disposal problems associated with mining and surface processing and minimizes the use of scarce water supplies.

In theory the process will work this way: A specially constructed heating-pumping unit would be lowered into a small vertical hole drilled where there are known oil shale or other hydrocarbon-bearing deposits. This unit provides not only a means to transmit electrical energy into the subsurface deposit, but also serves as a conduit for bringing liquid hydrocarbons to the surface. This energy does not radiate into the atmosphere.

Successful demonstration of the process would open up for economic production the vast U.S. reserves of shale oil, estimated to be in excess of one trillion barrels—equivalent to more than 150 years' supply at the current rate of consumption.

The participating companies have extensive experience in the technologies necessary to the successful development of the new process. Raytheon, which has been active in a variety of energy programs and in commercial and government electronics, started the in situ radio—frequency work seven years ago at its equipment development laboratories in Wayland, Massachusetts. The Badger Company has been engaged for many years in the design, engineering, and construction of petroleum and synthetic fuel processing facilities. Texaco is a leader in the technology of oil producing and refining, and has been actively involved in the development of oil shale and tar sands technology for several years.

U.S. patents relating to the work have been issued, and other patent applications are pending.

### The Tosco Sand Wash Project

The Uintah Basin of Utah, about 35 miles south of Vernal, is the site of the Tosco Corporation's Sand Wash Project, which will yield 47,000 barrels of upgraded shale oil per day when commercial production is reached.

The Sand Wash complex at commercial scale will use TOSCO II surface retorts to process daily 66,000 tons of 35 gallon per ton shale. The process will require 9,000 acre-feet of water annually.

The firm has spent $4.3 million through 1980 on its project.

TOSCO received approval in January 1976 from the Utah Board of State Lands to unitize 29 state oil shale leases covering 14,688 acres and is now proceeding with project planning, design and permit acquisitions for its commercial-scale facility. Major permits to be acquired for Sand Wash are mine plan and surface rehabilitation approval from the Utah State Division of Oil; rights of way permits for production pipelines, power lines, conveyors, roads and water supplies, all from the Bureau of Land Management; air quality permits from the state and from the Environmental

Protection Agency; and state and EPA approvals for any water discharge or reinjection.

The most likely product from Sand Wash is low sulfur hydrotreated shale oil, which will be shipped by pipeline to either Salt Lake City or to Midwest markets via Rangely. Local markets are preferred. Marketing outside of the region is heavily dependent upon the availability of exchanges.

Actions leading to a site specific Environmental Impact Statement have been initiated and detailed environmental impact analysis is underway. The plant, similar in design to the Colony plant in Colorado, should have no discharge of liquid effluents. However, there may be temporary disposal of excess mine water. Air effluent discharges will comply with all current federal and state regulations. Sulfur compounds and ammonia generated will be routed to processing units for recovery of elemental sulfur and ammonia for sale as by-products.

The disposal of spent shale will cause land disturbance at the project site. Topsoil will be removed ahead of filling and stockpiled for reuse in land reclamation. Revegetation of all disturbed areas is planned to re-establish wildlife habitat disturbed during plant construction. No impact on the agriculture of the largely arid region is anticipated.

Construction of a water impoundment facility on the White River to serve oil shale development in the Uintah Basin is the subject of an Environmental Impact Statement now being processed.

The current estimated population of Uintah County is 20,200 based on the 1980 U.S. census household count. Vernal, the largest city, has a population of about 6,500. About 10,000 new permanent residents will move into the region as a result of TOSCO's Sand Wash project.

## Union Oil Company of California

Union Oil Company has been involved in the oil shale industry for more than 50 years since it first began acquiring properties in the Parachute Creek area of Western Colorado's Garfield County in the early 1920's. It owns outright some 20,000 acres of oil shale lands containing reserves adequate to produce 50,000 barrels per day of shale oil for at least 90 years with today's technology. In addition, it owns more than 10,000 contiguous acres of valley and ranch lands acquired for support of shale retorting operations. Union realized long ago that shale development would require substantial water, and over the years, had taken steps to assure water supplies which are now estimated to be adequate for full development of its property at the rate of 150,000 barrels of shale oil per day.

Considerable research and other work necessary to the development of the resource has been accomplished over the years. In 1943, Union began active development of its own unique oil shale retorting process, first building and operating a small two-ton per day pilot retort at its Los Angeles refinery. Later, in the early 1950's, a 30-ton per day research unit was built and tested. This retort was based on the rock pump upflow principle, with heat for retorting supplied by combustion of the carbonaceous residue left on the shale after retorting.

Based on this early pilot plant experience, Union designed and constructed a semi-works plant in Colorado in early 1957 and operated it for a year and a half. During this time sufficient data were obtained to permit engineering design and economic evaluation of a proposed commercial scale plant.

The semi-works plant processed up to 1,200 tons of ore per day and produced as much as 800 barrels of shale oil per day. Over 13,000 barrels of this shale oil were processed successfully into gasoline and other products at a Colorado refinery. Subsequent research and development work led to further improvements and resulted in today's retort concept referred to as the B Mode of the Union Upflow Retorting Process. This externally heated retort has retained the rock pump upflow principle. The unique advantage of the upflow retort is that the retorting zone is at the top of the bed, which assures high yield, high quality oil, and trouble-free operation. The Union B retort produces offgas with a Btu content of about 80 percent that of natural gas.

In the belief that adequate financial assistance can be negotiated with the federal government, in the fall of 1980 Union began construction on the first phase of its Parachute Creek Shale Oil Project, which will produce 50,000 barrels per day of shale oil when completed in 1987.

The first phase of the project is a 12,500 ton-per-day room and pillar mine and surface retort which will produce 10,000 barrels of raw shale oil per day. All necessary local, state and federal government permits have been obtained for this first mine and retort. Completion of this first unit is scheduled for 1983.

Union also plans to construct a 10,000 barrel per day raw shale oil upgrading facility on its Parachute Creek property with completion also scheduled for 1983. The upgrading facility will convert the raw shale oil into a pipeline quality syncrude usable as feed stock in conventional refineries.

After the first units have operated long enough to confirm technical, economic and environmental characteristics of the project, Union plans to open four more mines and build four more retorts and additional upgrading facilities to bring capacity to 50,000 barrels per day by 1987. The total cost of the 50,000 barrel-per-day project, including upgrading facilities will be on the order of $2 billion (in 1980 dollars).

Shale for the first 10,000 barrel-per-day module, which will be mined from the rich Mahogany zone of the Green River Formation, is expected to yield 35 gallons of shale oil per ton. Located about 1,000 feet above the valley floor, the underground mine will open at a portal on the south side of Long Ridge, which forms the north wall of the East Fork of Parachute Creek. Conventional room and pillar mining will be employed.

The initial mining equipment will consist of front end loaders and/or hydraulic shovels loading into large off-highway type trucks, which will carry the ore to the crushing area.

The crushed ore will then move to the retort feed chute and into the solids feeder or rock pump. A 10-foot diameter piston then forces the ore upward into the retort. Hot recycle gas is introduced into the top of the retort releasing kerogen (raw shale oil) contained in the shale in liquid product which flows down through the cool incoming shale and the balance is carried from the retort by the recycle gases.

The retorted shale, with all of the oil removed, is forced up and over the edge of the retort cone and then falls by gravity through a cooling system for removal to the disposal area. It will drop through an enclosed chute into the East Fork valley floor deposit area where it will be spread, compacted, contoured and vegetated with native plants to blend into the surrounding landscape. Research over the past several years has proven that it is possible to revegetate retorted shale successfully.

The raw shale oil will be upgraded at a facility three miles northwest of the town of Parachute. Sulfur, nitrogen and other impurities will be removed leaving a syncrude superior to most grades of conventional crude oil. Jet and diesel fuels also may be produced at this facility.

Water for the initial retort and upgrading facility will be supplied from wells drilled on Union Oil's property. As future retorts and process facilities are added, water also will be drawn from the Colorado River, where Union has long-established water rights.

During the construction phase of the project, employment will average 2,000 persons for the years 1981 through 1987, with a peak of 4,700 in 1986. A permanent operating work force of 1,100 people will be reached in 1988.

Recognizing the socioeconomic impact of a project of this magnitude in a sparsely populated area of Colorado, Union has worked closely with local and county officials to alleviate problems that are foreseen. Union has agreed to provide front end financing for a middle school for 225 students. Union also will prepay to the town of Parachute, $600,000 for expanded sewer and water service. Union will build a camp for single employees on its property for up to 1,400 construction workers. Up to 645 housing units (rental and for sale) will be built in Rifle and Parachute and direct grants will be made for additional county and city law enforcement and fire protection service.

### White River Shale Project

The White River Shale Project (WRSP) is involved with the development of two federal prototype leases—Tracts Ua and Ub. Both of the Utah leases went into effect June 1, 1974. Tracts Ua and Ub will be developed jointly by Phillips Petroleum Company, Sohio Shale Oil Company and Sunoco Energy Development Company.

In 1977, the owners were granted a preliminary injunction that indefinitely suspended the terms and obligations of the leases. This injunction was based on conflicting claims on the title and is still in effect.

Once the injunction is lifted, the White River Shale Project plans call for a staged development in three phases. Phase I comprises mining and retorting designed to generate specific information on ore body characteristics and information on the retorting process itself. Up to 30,000 tons of shale per day will be mined in this phase. This will result in an oil production of approximately 15,000 barrels per day by the end of 1984.

WRSP will use proven room and pillar techniques to mine the raw oil shale. Raw shale will be crushed and screened to produce the required size ranges. After crushing, the raw shale will be processed in surface retorts.

Phases II and III cover the expansion of shale oil production to a level of 50,000 to 100,000 barrels per day, respectively. Construction of Phase II is planned to begin in 1986, with operation of the 100,000 barrel per day commercial facility expected to begin in 1993. Included in all phases are the support facilities needed at each stage of development, including facilities to supply water, power and access requirements.

Manpower requirements for the operations will build steadily from an initial need of approximately 200 people to 3,200 people in 1993. Thus, the oil shale operations rely on a very stable workforce.

The construction workforce, however, is of a temporary nature. Phase I's construction manpower requirements peak at 1,830 workers in 1983 and Phase III's construction requirements peak at approximately 4,000 workers in 1988-1989.

As a result of the differences in the nature of the construction workers and the operations workforce, different approaches will be taken to minimize the impact these workers may have upon the local communities. Since the individual construction workers will remain for only short periods of time, a construction camp will be established on-site. This housing will be temporary in nature.

Because the operations personnel will be a permanent workforce, every attempt will be made to assimilate these people into the existing local communities. This will require close coordination between WRSP and the local communities to develop a suitable plan.

Any development of tracts Ua and Ub is unlikely to occur, however, until the legal questions concerning the ownership of the leased lands is resolved. WRSP is currently working towards such a resolution.

## Other Technologies

In addition to these identified projects, other surface retorting process schemes were also under investigation. These included Superior and Dravo Circular Grate Retorts; and Lurgi Retorts.

The Superior and Dravo Circular Grate retorts could be operated either as direct or indirect retorts. In these retorts, the shale is loaded onto a moving circular grate conveyor contained in a hollow doughnut-shaped vessel. The shale travels around the circle and is discharged. There are five zones in the retort:

The shale feed zone
The shale heating or retorting zone
The carbon burning zone
A cooling, condensing zone
The spent shale ash discharge zone

Vapor and gases are channeled through these various zones or compartments by internal ducts and baffles. A pilot scale device representative of the Superior retort has been tested at a 250 TPD rate.

In the Lurgi process, the heat transfer medium is the spent shale itself. A circulating stream of hot spent shale ash is mixed with fresh unretorted shale in the screw mixer. The retorting reaction begins immediately and is completed in the surge bin below. The carbon values on the spent shale are captured in the Lurgi process by burning in the lift pipe. This portion of the process is fluidized by the combustion air-very comparable to catalytic cracking technology. Spent shale ash is separated from the flue gas leaving the lift pipe.

The Lurgi process has the advantage of both *indirect* and *direct* retorts since the *indirect* undiluted gases are available for sale or used elsewhere and the heat is supplied by direct combustion of the carbon residue on the spent shale.

Lurgi retort equipment components have been demonstrated in commercial size in other applications such as the devolatilization of coal.

In addition to these technologies currently under development, there were other experimental methods, which may prove useful in the future. Some are retorts, some involve other techniques:

> In situ laser processes such as some which were then under development in Israel.
>
> Radio frequency technology is in the early experimental stage and operates on principles similar to microwave ovens.
>
> Possible use of biotechnology for surface retorting. Bio-separation of oil from shale appears to be well into the future.
>
> Fluid bed retorting in surface retorts. This could be a second generation technology as briefly discussed by Chevron above.

Some of these experimental technologies will probably be developed, and certainly improvements will be made in existing technologies as projects proceed to commercial development."

It can be seen from reading the above that during the years around 1980, considerable optimism existed about the potential of an oil shale industry. Also, hedging of bets was very prevalent. There existed considerable uncertainty about government participation at all levels, political reverberations from unknown influence of environmental and other powerful lobbies and, likely most significant to the energy companies, the effect of double digit inflation and money availability. It is interesting to me that the one venture that expressed the most realism in 1981, e.g., $2 billion capital cost and the need for personnel camps and apartments in nearby towns, etc., was Union Oil and indeed was the only venture that eventually proceeded with construction and operation and produced a marketable quantity of shale oil. This occurred, notwithstanding their use of one of the most complex processing technologies being proposed, with the careful employment of an ironclad agreement with the Federal government, which provided a substantial safety net under their financial risk.

# CHAPTER 6

# Construction Begins at C-b

THE FIRST GENERAL contractor chosen by the C-b team was the Ralph M. Parsons Company of Pasadena, California. Actually, we had taken bids and chosen Parsons while Shell was the operating partner. Their business developer, Alan Silberberg, made us an exceptionally good offer and provided us with good quality personnel. When Oxy became the program operator, the Parson's contract was extended, for project continuity and because Parsons and Oxy had a history of working together on the Logan Wash MIS works. As is customary in this type of project, we set up a project team in Pasadena where the loci of Parson's technical people were stationed. Parsons and we also set up a field management organization on the site. There was already a good start for this, as the environmental baseline data development had required many field people with substantial amounts of construction equipment.

The site was cleared according to the approved DDP, in other words, designed to make a minimum impact on the terrain, air and water. All topsoil was carefully stockpiled in segregated areas for subsequent use in revegetation and reclamation of disturbed areas. Construction office buildings were erected, access roads developed, water, electricity, telephone, and other utilities provided, again all under the watchful eye of the AOSS who had the primary responsibility to the taxpayers for our compliance with the lease and DDP. This validated again to me the early decision to work in a partnership with the AOSS rather than in an adversarial relationship.

Our first major effort was to sink three shafts to access the shale deposits. Part of the original thinking behind the prototype leasing program was to test differing geological and geographical circumstances, i.e., C-a was supposed to test close to the surface deposits to be mined by open pit and C-b was supposed to represent a deep deposit requiring large shafts and underground tunnels to mine and bring the shale rock in great quantities for removal of the oil in retorts located on the surface. As noted in the previous chapter, the best-laid plans . . . often go astray and neither of these presumed mining/process schemes were being advanced.

Also, I know that the approved DDP as described in the previous chapter said that C-b would be sinking 3, 4, or 5 shafts depending on which page you were reading. However, it is characteristic of the developing oil shale industry that you had to be flexible. Things were so dynamic and parameters changing so quickly from day to day, that a dated program was needed to be able to state with any degree of assurance just where each project was. Again, I can't emphasize enough the impact that inflation was having on our efforts. It is no wonder that oil company management was becoming snakebit, with we engineers coming in with a new and significantly higher estimate every time we were asked. Also government officials had a right to be skeptical of our forecasts, because they were never the same. And certainly, the environmental folks had great reason to cast doubts on where the industry was really wanting to go and our honest intentions because of the moving targets we continued to give them.

A fascinating discussion came to light about this time. The US Department of Energy was one of the government agencies getting blind-sided by industry estimates. They contracted with the RAND

Corporation to study and analyze the situation. Their conclusions appealed to all and especially bankers (who just might have been the Federal government in this case). Among other points, they concluded that, in line with Cyril Northcote Parkinson, the British historian and author who formulated "Parkinson's Law," the satiric dictum that "Work expands to fill the time available for its completion";

> "Optimism tends to expand to fill the scope available for its exercise. This Parkinsonian pronouncement was interpreted as follows: "If it is not possible to penalize inaccurate estimation, or if inaccuracy is actually rewarded, the scope for optimism increases. The ability to penalize for inaccurate estimation is largely a function of how far into the future a technology's commercial readiness lies. In addition, the ability to bestow appropriate reward for accurate estimation requires being able to distinguish among deliberate misestimation and sloppiness, the effects of poor management, exogenous factors, and inadequacies in the estimating state of the art for new technologies." [vi]

Other than the mysterious "exogenous factors", I would have to at least acknowledge the existence of those condemnatory qualities in our business during this period. This was indeed a period of high intensity competition and salesmanship.

To further get our attention, the RAND folks concluded:

> "The assumptions that planners make about the accuracy and uncertainty of their capital cost estimates are frequently unrealistic.
> Estimates that are made for projects that use commercially unproven technologies not only are characteristically biased low, but **also are so uncertain that they cannot be relied upon at all.** (My emphasis added.) Despite their notoriety, the major villains in cost growth **are not uncontrollable or external influences such as inflation** (More of my emphasis) or "scope changes." Most of this bias and uncertainty result from low levels of process and project understanding, particularly for new technologies." [vii]

These folks were playing for keeps. However, salesmanship cannot be separated from even the lofty research and development ivory towers. It probably should be acknowledged that at least some of the group at RAND were selling a statistical cost prediction model which would presumably greatly ease the weighty burdens that the folks at the DOE were experiencing at the time. Again, I cannot expressly deny all of their suggestions, but I would plead that unpredictable price and wage increases upwards of 10% per year were causing major perturbations in the estimating business. Scope changes from the industry research and development organizations based upon their rather frantic on-going testing of proposed technologies were also adding to the confusion. It was a splendid time for lots of finger pointing and limited acceptance of responsibility.

Right in the middle of all this unsettled confusion, we, the C-b project team, were still charged with designing and constructing appropriate facilities to move the project forward under the constraints of the approved schedule which at the time was the one noted in the DDP write-up of Chapter 5. So we released Parsons to obtain bids to build headframes and sink four (whoops, didn't I just say three, or maybe five shafts? So it goes.) large shafts, along with necessary electric power supplies and dams, etc., as needed to handle water pumped as shafts were sunk. In addition, engineering design for the large hoists and motors required to get men and materials into the mine after the shafts were sunk was initiated.

To give some idea of the magnitude of expenditures being laid out by the partners (a steadily changing roster of companies) here are financial data for the four years 1978 to 1981 inclusive:

| Year | Total Expenditures |
|------|-------------------|
| 1978 | $14,045,000 |
| 1979 | $29,971,500 |
| 1980 | $50,512,000 |
| 1981 | $76,534,000 |

The 1981 costs, as an example, were further broken down into the following categories:

### 1981 C-b Expenditures
(Thousand of Dollars)

| | | | |
|---|---|---|---|
| *Field Construction* | | | $47,134 |
| *Engineering Cost* | | | $10,078 |
| *Tract Operation &* | | | $2,491 |
| *Maintenance* | | | |
| *Oil Upgrading* | | | $700 |
| *Environmental* | | | $944 |
| *Other Programs* | Housing | 232 | |
| | Community Relations | 187 | |
| | Busing | 775 | |
| | Insurance, Bonding and Property Tax | 803 | |
| | Land | 371 | |
| | Other | 47 | |
| | Total Other Programs | | $2,415 |
| *General & Administrative* | Environmental Staff | 780 | |
| | All Other Staff | 8,481 | |
| | Expenses | 2,967 | |
| | Overhead | 543 | |
| | Total | | $12,771 |
| **TOTAL PROJECT** | | | **$76,534** |

# CHAPTER 7

# Occidental Oil Shale Company

A S MENTIONED, OCCIDENTAL became an active participant in Tract C-b when Shell left Ashland as the sole participant in late 1976. Oxy had been working on their MIS process at Logan Wash outside of Debeque, Colorado since 1972. The Oxy shale oil business was a pet project of the charismatic chairman of Occidental Petroleum Corporation, Dr. Armand Hammer. He was a charming and unique individual and I have never regretted going to work for him. Because of his intense personal interest in oil shale development, we were blessed with his presence way more than a minor subsidiary of the good sized parent corporation would normally receive. He participated in meetings on many details of the process and wanted to know exactly what was happening during each stage of the research and development. An interesting personal experience of mine was sitting next to the doctor during an afternoon meeting in our crowded quarters on the outskirts of Grand Junction, Colorado. My secretary came into the meeting and gave me a note that said that President Carter was holding on my phone for Dr. Hammer. Now the good doctor was in his eighties at this time and was in the habit of napping briefly after lunch. So I had to nudge the man and then repeat the message to him a couple of times. He jumped up—eighty or not, he was very spry— and went to talk to the President of the United States. It seemed that President Carter wished for Dr. Hammer to use his contacts in the USSR to facilitate some transaction between the super powers. So the doctor started placing calls all around the world through our somewhat antiquated phone system. Our receptionist at the time, Karen Wyman, was one of the sweetest voiced young ladies you'll ever meet. It was wonderful hearing her deal with the many foreign operators and get all the connections appropriately made and both the doctor and the senior leaders of the US and the USSR satisfied.

Doctor Hammer used his well known name and presence to sell shale oil development to government officials at every available opportunity. One such discussion impressively displays the doctor's message on our product and project. He commissioned the production of a glossy 40 page booklet in 1979 to address the propositions of the President's National Energy Plan of April, 1977 which said in part, " . . . Unless the U.S. makes a timely adjustment before world oil becomes very scarce and very expensive in the mid 1980's, the nation's economic security and the American way of life will be gravely endangered. The steps the U.S. must take *now* are small compared to the drastic measures that will be needed if the U.S. does nothing until it is too late." Also quoted as justification for the booklet and the Oxy development was a statement by the then Secretary of Air Force, John C. Stetson, at an International Synthetic Fuels Conference at Menlo Park, California on March 8, 1979 as follow, "The Air Force (is) aggressively developing shale oil fuel specifications and conducting refining and engine performance cost studies . . . we anticipate delivery of worthwhile quantities of synthetic fuels to defense perhaps as early as the mid 1980's."

In a preface to the booklet which outlined development of the resource in an environmentally acceptable way, the history and economics of the modified in-situ process and a justification for government support of the oil shale, Dr. Hammer was at his best in describing his fervent belief in the viability of the development of this tremendous energy reserve. His message was as follows:

"Fellow Americans:

The current and future availability of crude oil is the nation's most pressing energy problem. No other fuel source or known technology can be expected to power our cars, trucks, tractors, planes and ships, nor our machines of defense, within this century.

We must develop alternative liquid fuels. One such alternative fuel—shale oil—is ready now, and could be providing us with significant quantities of liquid fuels by the mid-1980's. This nation has vast quantities of oil shale. We now have the technology for recovering oil from this domestic resource. What we have so far lacked is the determination to commit the resources and provide the incentives necessary to develop a national shale oil capability

It is our perception that few people know about oil shale and the technology for extracting oil from it. It is our purpose to provide this document to better inform the public and government about this domestic resource and our nation's need to utilize it.

Let me put our present situation into perspective. The escalating price and growing shortage of petroleum are beginning to influence our life styles and employment patterns, and to compromise our standard of living. Our dependence on foreign sources of crude oil is increasing this nation's susceptibility to political events in other parts of the world and is jeopardizing our national defense. Today we import almost one-half of our oil from overseas, consuming one-fourth of the entire world's available oil exports. The cost of these foreign purchases will exceed $60 billion in 1979. This reliance on foreign fuels is eroding the U.S. dollar in world markets, adding further to our already spiraling domestic inflation.

Since the Oil Embargo of 1973, this nation has done little to expand domestic production of liquid fuels or to develop alternative energy sources. Charles Schultz, Chairman of the President's Joint Council of Economic Advisors, stated this most directly in his

testimony before the Joint Economic Committee, Subcommittee
on Energy on April 25,1979:

' . . . our controls on oil discourage the development and
production of alternative energy sources. Every barrel of oil
equivalent that is produced from non-oil sources, like synthetic
fuels, saves the nation some $17-$18 in reduced oil import costs.
But we force producers of alternatives to sell into a market where
they must compete with oil whose price is controlled at the lower
average price. Thus we discourage investment in such alternative
sources, at a direct loss to the nation. *Incredibly under the current
control system, we pay OPEC more for oil than we are willing to pay
Americans who produce oil substitutes. It would be hard to design a
system more carefully calculated to encourage oil imports and slow down
the development of alternative domestic sources:'* (Emphasis supplied
in original)

A number of technologies hold promise for the future—solar,
geothermal, nuclear fission and fusion. But none of these is capable
of making major contributions to supplement or replace our
dwindling petroleum supplies during the next ten to twenty years.
You cannot run an automobile on any of these energy forms.

Oil from the rich oil shale deposits in the Rocky Mountains
can provide significant quantities of domestic liquid fuels at a
reasonable price over the near term. The technology for the large-
scale production of shale oil has been demonstrated. We know
how to do it. And we believe we can do it in an economic and
environmentally acceptable manner. Over 500,000 barrels of oil
from shale have been recovered by other private companies and us.
This shale oil is compatible with our existing fuels. As-produced
shale oil without upgrading has been burned directly in utility
boilers. Shale oil has been refined to provide gasoline and has been
used successfully as a jet aircraft fuel by the Navy.

Occidental Petroleum Corporation has spent over $100
million of our own monies to develop our shale oil extraction
process. We are planning to initiate production on a commercial
scale on our 5,000 acre Federal C-b tract in Colorado. We have

licensed Rio Blanco Oil Shale Company, a partnership of Gulf Oil and Standard Oil (Indiana), to use our technology in their development program. We have just entered into an agreement with Tenneco to sell them a half interest in the C-b tract in return for $110 million reimbursement to us attributable to work done on this tract after April 1, 1979 by Occidental. Thereafter each partner will pay equally for future work. This is subject to a mutually satisfactory partnership agreement and approval by the Boards of Directors of both companies. Recently, we signed an agreement with Morocco to investigate the possibility of using Occidental oil shale technology to develop a 50,000 barrel per day facility in that country. This facility if built, could provide Morocco with virtual energy independence. Our optimism is shared by other firms that have similarly invested their own private funds and developed their own technologies to recover shale oil.

The national interest demands that this prodigious resource be tapped as quickly as possible. Yet the sheer magnitude of the investment facing the interested companies—about $1 billion to produce 50,000 barrels of oil per day—demands a cautious and businesslike pace. This pace could be accelerated by firm government action and a demonstrated national commitment to the production of shale oil as set forth in the President's Address on July 15th.

In testimony before the Congress, I have outlined a plan to foster the development of a two-million-barrel-per-day shale oil industry by the year 1990. I have offered the government free license for defense purposes to Occidental's Modified In Situ oil shale process, and we stand ready to license other commercial firms for a reasonable fee.

The Oil Embargo of 1973 warned us of the danger we face of having the major supply of our imported oil cut off at any time. We know that oil worldwide is a depleting asset and will decline significantly over the next 10 to 20 years. Six precious years have passed since the Embargo. The need is here, now, today, to begin the very large-scale production of oil substitutes if this nation's economy is to survive. In this document, I have set forth a blueprint for building energy independence from imported oil. Remember, you cannot drill a dry-hole in shale; the oil is there. If we can put an American on the moon, surely we can get this oil out in an environmentally acceptable manner at a reasonable cost. I am confident we can do the job, and I am proud that Occidental is leading the way."

# CHAPTER 8

# Logan Wash Operations

W HEN I WAS hired by Oxy, a significant part of my initial responsibility was to keep the research and development operations at Logan Wash intact and productive. Occidental initiated efforts to begin experimental Modified In Situ operations in 1968. Three years of effort were required to obtain land and environmental permits for the field tests. A tract of approximately 4000 acres at Logan Wash near Debeque, Colorado was purchased from private sources in 1972. The site was estimated to contain an estimated 750 million barrels of shale oil in medium grade deposits, 250 feet thick averaging about 17 gallons of oil per ton of shale. Mining operations began in 1972, and the first retort was fired up the following year. Operations continued through a series of progressively larger retorts. Retort 6, one of those constructed in my time there had one-hundred times the volume of the first experimental unit and was advertised to be of an appropriate size for commercial operations.

The initial three retorts were located off a single horizontal mine opening. Retort lE ("Experimental") was mined in the form of a small room with a vertical cylindrical rise providing the initial void volume. The rubble pile was 72 feet high. Start-up was delayed by ignition, temperature and pressure measurement, and control problems in the retort. Satisfactory start-up was achieved in July 1973. The retort operated successfully producing over 1200 barrels of oil. For the first experiment, this was considered a high recovery rate.

In Retort 2E, the void volume was reduced and the blast pattern modified. The retort depth was increased 22 feet, extending the retort into an area of lower grade shale. Fired in 1974, Retort 2E produced 1400 barrels of oil.

Retort 3E tested an entirely different retort design, with different operating characteristics. Although some new problems were encountered, Retort 3E produced 1600 barrels of shale oil.

Following completion of the first three retorts, operations were transferred to a new large-scale development mine. Here the parameters were designed so that the mine could later be converted to commercial operation. Retort 4 was the first commercial size unit. It was fifty times larger than the first retorts and designed to permit evaluation of factors not critical with smaller retorts. Specific topics evaluated in the scale-up included geologic and rock mechanic factors, enlargement of the blasting pattern, and retort flow control over a larger cross-sectional area. Included within the retort was one zone of barren rock, and another zone containing rich, 35 gallon per ton shale. Operation of Retort 4 produced some 32,000 barrels of oil. This was somewhat less than the full potential. Difficulties were traced to geologic conditions, which resulted in inadequate rubblization of the ore.

In Retort 5, the geologic conditions that complicated the previous blast were overcome by several design changes. The blasting pattern was modified and the volume mined was reduced in an effort to lower costs and enhance the possibility of utilizing low grade shale. From diagnostic tests prior to ignition, it was determined that the available void volume was not distributed uniformly throughout the retort. This was later confirmed when channeling occurred and the oil yield was found to be lower than expected. Corrective measures were partially successful in increasing oil yield. A total of 11,300 barrels of oil was recovered from this retort and samples refined at Chevron's Salt Lake City refinery.

Retort 6 was a scale-up of the successful Retort 3E design. It measured 162 x 162 feet at the base and 269 feet tall—half an acre in cross-sectional area and as high as a 30-story building. The void volume in the retort was created by mining of a horizontal room to provide more uniform permeability in the rubble zone and hence hopefully better oil recovery rates. Rubblized on March 25, 1978, Retort 6 was ignited on August 28,1978. Operating conditions were upset soon after start-up by a partial collapse of the retort roof (sill pillar slumping into the retort). However, corrective actions were taken and over 50,000 barrels of shale oil were produced from this retort of which over 46,000 barrels were recovered. This represented 40 percent of the oil in place. Analysis of retort performance indicated that rubblization was satisfactory throughout the entire retort. Had the sill pillar remained intact, total production from Retort 6 could have been as much as approximately 75,000 barrels of oil.

I have always felt that our experience with Retort 6 may have revealed intentions of fate that I had been searching for about oil shale development. When the sill pillar collapsed, the combustion zone moved backwards to the air supply level of the mine. We quickly, and in my opinion, professionally sealed off much off the air supply level. The combustion zone then was forced to a vertical raise shaft to the surface of the mine. Copious quantities of smoke poured out of the raise and caused us to advise surrounding communities of the problem. After long hours of remedial action and seeing the situation stabilize although smoke was still moving into the valleys, we decided to go home for some needed rest before tackling the next phase of remediation. We of course left a well trained and equipped crew on duty at the mine. Early the next morning as we drove back to Debeque, we were all dreading the sight of barely controlled smoke surging from where it wasn't supposed to be. Lo and behold, as we drove around the critical curve in the highway, we couldn't believe our eyes. Instead of black smoke, we saw a beautiful Colorado morning. It turned out that the raise through which the hot gas was rising had caught on fire enough to cause the sides to collapse and naturally plug the several feet in diameter raise. The crisis was essentially over. With a minimum of effort, we were able to stabilize the mine and continue the recovery of the oil from the retort. So, as I say, it seems to me that it may very well have been mystic intervention which fixed our problem and may be some sort of sign that modified in-situ shale oil production is destined to happen and the promise of the rainbow was valid for our enterprise.

My granddaughter and I observed a further example of possibly sanctified intervention into human interference with the works of nature as we drove along the San Rafael Swell in central Utah. There, in some of the most beautiful and lonesome country I know, a New Yorker had stopped by the side of the freeway, and was driving a golf ball into one of the remote canyons which cut across the plateau. He undoubtedly was going to brag how he had hit a drive 500 yards on his vacation. Just as he swung and as we drove by, lightning struck close by. The golfer ran for his life back to his car. We thoroughly enjoyed the spectacle of Nature protecting her property with a well-timed warning. As I say, these experiences have convinced me that shale oil production has been given a celestial pass to proceed.

Occidental proceeded with plans to construct Retorts 7 and 8 incorporating improvements from the Retort 6 design.

By 1980, Occidental had been awarded sixty-eight patents covering its Modified In Situ technology; an additional twenty-two applications had been allowed and thirty-four more were pending.

# CHAPTER 9

# On with C-b

IT IS PROBABLY appropriate to update the ownership/operator status of the C-b project. We last left the discussion with Oxy becoming a 50/50 partner with Ashland Oil, taking Shell's share. There was no money exchange at that time, Oxy contributed their modified in-situ technology as developed at the Logan Wash site. Subsequently, Oxy increased their share to 75% and in February 1979, Ashland withdrew completely from the project, leaving Oxy with a 100% interest in the Tract C-b lease. However, negotiations were underway with Tenneco Oil Company who acquired a 50% interest in the lease later in 1979. Tenneco paid Oxy $100 million, which seemed great to me as an Oxy employee, since Oxy had hardly put any of their own money into C-b. Most of Tenneco's money was to be spent before Oxy had to share costs 50/50 and Tenneco management had committed the funds. From my perspective, $100 million would move us quite a ways down the road to development.

At that time, the Oxy/Tenneco partnership was renamed the

Cathedral Bluffs Shale Oil Company for a distinctive topographical feature nearby and logically, because the Cathedral Bluffs initials coincided with the government C-b designation. Oxy was the designated operator of the partnership at that time and as noted above, money was readily available, so it was a grand time for me and the rest of the projecteers.

It was an extremely exciting time. We had a big organization with several major contractors hired, including Bechtel, Fluor-Daniel, Brown & Root, and Dravo. It was my job to coordinate the work going on in the contractor's offices with that going on in the field. I spent much of my time on the road checking out the project teams in the contractor's offices and then going to the field to check construction progress. I had to report to management monthly, but once we'd get an annual budget approved—which was always a struggle of tactics and politics—it was mostly a matter of reporting progress against expenditures.

The shaft sinking was exciting. Our contractor was the very professional Gilbert Corporation, an open-shop arm of the Peter Kiewit group. One shaft was in a very wet zone and it was like working in a rain forest down there. I was always amazed that they could progress as well as they did. The service shaft was one of the largest being sunk in the world—a significant accomplishment.

Before the shafts could be sunk, headframes needed to be constructed. These structures were very large in themselves. Construction of the 29-foot diameter Production Shaft was begun in February, 1978 when it was "collared in" to approximately 70-foot depth by conventional excavation methods. The 313 foot headframe was "slipformed" in just 26 days during September and early October, 1978. Slipforming is a method of continuous construction in which the form is slipped or jacked-up as the concrete is poured in place. Rebar is placed ahead of the advancing form. After slip forming, steel beams and floors were installed and collar floors completed; after which the roof and lower power floor beams were set. Temporary sinking and Galloway hoists were installed in 1978. Both were housed in metal buildings erected near the shaft. The Production Shaft was to serve as the main "mined-rock" hoisting facility during commercial operation.

In April 1981 construction to equip the headframe for full-scale mining began. Feeder extensions, structures to house the two feeder conveyors that would transport mined material away from the headframe, were erected. A permanent 60-ton bridge crane, also used in aiding construction, was installed just below the headframe ceiling. Three concrete floors for supporting hoisting equipment were erected, the Upper Power Floor and the Lower and Upper Hoist Floors. Electrical equipment, such as switchgear, transformers and speed control thyristors, was set in place. Electrical conduit and cable trays were installed, wiring was pulled, and lighting fixtures installed. Preparation for installation of the two large skip hoists got underway; and by year's end, the sole plates on which they would rest were nearly completely installed.

The 34-foot diameter Service Shaft was to be used for hoisting both men and equipment and as a ventilation air intake. Construction of the shaft commenced in February, 1978. The collar and headframe foundations were completed at a depth of 65 feet in May, 1978. An air inlet or "air tunnel" to the Service Shaft was completed during August, 1978; it enters the Service Shaft some 100 feet below grade. Slipforming of the headframe tower took place in a 10-day period during August 1978. Facilities for sinking were installed as follows: the manloading and collar floor was completed and beams set in the roof, upper power floor, lower power floor and dump chute. The dump chute and collar door were installed. The temporary sinking and service hoists were then installed and a metal building erected to enclose them. The shaft-sinking mechanical and electrical facilities were completed in 1979. This shaft contains the following floors (from top to bottom); hoist floors, upper power floor, lower power floor, collar floor (ground level) and man loading (sub-collar) floor.

In April 1981 construction began to equip the headframe to accommodate a full-scale mining operation. Initially, a 30-ton bridge crane was erected just below the headframe ceiling. This permanent crane became operational and was used for lifting material to erect the Hoist and Power Floors, where concrete was placed in July. Three hoists were installed, the 240-man Main Cage hoist and two smaller Auxiliary hoists including electrical hook-up. Concrete was placed for the floors and walls of the Mechanical/Electrical Room on the south side of the headframe and the West Airlock Building.

Construction of the 15-foot diameter Ventilation/Escape (V/E) Shaft commenced in May 1978; the collar and headframe foundation was completed that June. The structural steel headframe was erected and a siding and dump chute installation were completed. A metal building to house the shaft sinker's shop and dry room was erected in 1978. Construction of the V/E Shaft hoist house was begun during July, 1978; the building itself was completed in December. Installation of mechanical/electrical facilities was completed in 1979.

In addition, electric power and switching facilities were needed and constructed. Primary power for site construction was provided by nine 1000 KW, 4160 V, natural-gas/diesel powered generators. Seven units were installed in 1978 and 1979 with two additional units installed in 1980. Power distribution was by 13.8 KV overhead power line to the shafts, batch plant, warehouse and office areas. Generators were placed on line as requirements demanded with at least one unit in stand-by mode at all times. In addition, there were four 250 KW, 440 V diesel-driven generators to provide emergency power to the batch plant and office/warehouse complex. Power for environmental monitoring stations, heliport, sewage treatment plant and security gate was furnished at a loading of 7.2 KV by the White River Electric Company via existing rural power lines. Site preparation for the switchyard and mine power substation was initiated in July, 1980. High-voltage equipment was installed at both areas by February, 1981. In April, 1981 erection of a new Meeker-to-C.B.13.8 KV power line got under way, and it was completed by October. This consisted of a single 138,000 volt, 22-mile transmission line. A building to house switchgear and transformers for the mine power substation was constructed in 1981. By year's end, most of the electrical equipment there was set in place, and electrical cable tray installation had begun. Also the line was successfully energized and tested.

Water for the batch plant operations and shower facilities was hauled via truck from a well on Piceance Creek. Potable water was hauled from Rifle.

A concrete batch plant produced 20,000 cubic yards of concrete a year for on-tract use. Bulk cement, sand, and aggregate, mixtures were transported to the site by truck, from Rifle or Meeker. Water for the batch plant operation was hauled from the well on Piceance Creek. Winterization of the batch plant included installation of bifold doors and steam heaters on the sand aggregate storage building.

The explosive storage (powder magazine) area was remotely located from areas of major activity. Approximately 98,600 lbs. of explosives were consumed in shaft sinking and mine level station development activities in 1981, a typical shaft development year. A total of approximately 550 blasts were detonated that year.

Significant water treatment facilities were necessary to properly handle water generated during the sinking of the shafts. The surface water facility was designed to dispose of excess mine water by direct discharge from two lower ponds or by sprinkler irrigation or by subsurface reinjection into the same general zones being dewatered in the shafts. The system was initiated in 1979 for direct discharge from the ponds into Little Gardenhire Gulch. In 1980 the sprinkler system was completed, tested and used throughout the summer. It was again in use during the summer of 1981. It consisted of a lateral distribution system on the ridge between Cottonwood and Sorghum Gulches. Reinjection commenced in May of 1981 and continued in almost constant use thenceforth.

The pH of the mine water was controlled to <9 at the ponds by the addition of sulfuric acid. A storage tank and associated piping delivered acid as needed to lower the pH of the water at each pond. The pH was monitored at the overflow between the ponds by a continuous pH meter. Grab samples were taken at other points in the ponds to assure that proper pH was being maintained. Suspended solids were settled out in the ponds by adding commercially available polymer flocculants with great success. This system consists of two 500-gallon mixing tanks and metering pumps, which fed the ponds. Settled solids were periodically dredged from the ponds. Gland seal pumps, located in the lower pond pump house, were continuously operated to provide cooling water to pumps which transferred mine water from the shafts to the ponds.

The sprinkler (land application) system was operated during summer months and pumped approximately 440 gpm through two moveable nozzles, for a total of 40 million gallons. The reinjection water treatment facility was tested and put into operation in 1981. Water was pumped from the ponds through an upflow sand filter. The filter lowered the turbidity of up to 700 gpm of water to a very low level. The filtered water was then reinjected with a centrifugal pump into existing wells. A total of 100 million gallons was reinjected during 1981. A total of 634 million gallons from the shafts was treated and disposed by construction usage, discharge to surface streams under NPDES permit, irrigation, or reinjection during 1981.

An on-site hydrology laboratory was equipped with all the necessary laboratory and safety equipment and supplies to ensure the proper preparation and testing of field water samples for pH, temperature, conductivity and dissolved oxygen as well as total dissolved solids and fluoride. When necessary, samples for additional water quality parameter determination were labeled, preserved and transported to Occidental's Grand Junction Laboratory for analysis.

Other on-site work during this typical construction year consisted of such items as drilling of two additional reinjection wells to approximately 1,600 feet depth; drilling of three new core holes for mineralogical and geological analysis; providing quality control efforts for all construction activities including civil, structural, mechanical, piping and electrical/ instrumentation work; installation of a fully equipped laboratory for concrete testing and contracting structural fill and soil analyses.

As mentioned, major engineering firms around the country were performing necessary design for the project. For example, Fluor Engineers and Constructors, Inc., Houston Division, performed engineering work that resulted in the following reports and estimates:

1.  Estimates of Costs of Aboveground Shale Retorting (AGR) Processes

2.  Process Design Package and Cost Estimate for MIS Surface Process Facilities for First Phase and Total Project.

3. Process Design Package and Cost Estimate for AGR Support Facilities for First Phase and Total Project.

4. Offgas Desulfurization Process Selection Report.

5. Process Flow Diagrams and Emission Data for the PSD Permit Application.

Additionally, the Ralph M. Parsons Company, Pasadena, California, evaluated sulfur control systems:

1. Evaluation and Comparison of Sulfur Dioxide Emission Control Systems,

2. Flue Gas Desulfurization Process Selection.

Stearns-Roger, Denver, prepared the Process Design Package and Capital and Operating Cost Estimate for a 12,800 tons per stream day (TPSD) Union B Retort.

Bechtel, Toronto, prepared a conceptual design package and cost estimates for surface material handling facilities to feed one AGR in the first phase, and six AGR's in the final phase of the project. Detailed engineering work was started on the first leg of the conveyor system from the Production Shaft to truck loadout bins. A crushing and screening test was also undertaken using C-b shale with the help of Bechtel personnel to determine the suitability of equipment to be utilized in the AGR feed preparation facilities. The test work was performed at the test facilities of four different vendors. Bechtel developed emission data from surface materials handling points for the PSD permit.

Development Engineering Incorporated prepared a report on a detailed laboratory examination of C-b shale. The purpose of this laboratory analysis was to determine the suitability of retorting C-b shale in the Paraho DH Retort.

Dames and Moore carried out a soils investigation in the surface process area and prepared a soils report.

Technology Management Inc. conducted studies on MIS oil shale emissions with and without de-emulsifiers.

Williams Brothers Engineering performed various studies on shale oil rheology, MIS shale oil hydraulics, and pipeline transportation of shale oil.

Resources Conservation Company of Seattle, Washington, prepared the following water related reports for C-b:

1. Cathedral Bluffs Oil Shale Facility Proposed Water Management System.

2. Interim Report, Fluoride Removal Project Engineering Evaluation

3. Optimize production of potable water from mine water.

SRI International of Menlo Park, California, performed a water treatability study of process condensate using the following technologies:

1. Steam Stripping, Single and Two Stage.

2. Fine Oil Removal.

3. Granular Activated Carbon Adsorption.

Brown and Caldwell of Walnut Creek, California, performed a biological water treatment study on process condensate.

Institute of Gas Technology conducted burner tests on simulated MIS offgas and issued a report entitled "Evaluation of Burners for Utilization of Offgas from Shale Gasification".

Arthur D. Little, Inc. conducted a marketing study for byproducts and raw materials.

Offsite facilities were required and constructed in nearby

communities. The headquarters facility at 751 Horizon Court in Grand Junction was put into use in January of 1979. We quickly outgrew even this expansive facility and moved the Accounting and Health, Safety, and Security Departments to the Crossroads Business Commons Building at 2764 Compass Drive (approximately one-half mile to the north of the headquarters facility) in 1981. We later expanded into offices across the street from the Horizon Court facility left vacant by Exxon when they folded their oil shale project.

An Oxy laboratory at 2372 G Road in Grand Junction was responsible for analyzing the various special and routine samples received from the Cathedral Bluff site. Analyses were run on samples of shale, gas, and water, as well as other materials such as sludges and minerals. Fisher assays, which told how much oil was in the shale rock, were also performed.

Forty acres were purchased west of Rifle for future rail siding, staging area, and product shipment. The project also used a rail siding adjacent to the Rifle railroad station for off loading bulk materials and equipment for construction.

An employee parking lot located behind the Rifle Gap Apartment units was paved and striped in 1981 at a cost of $256,000 and which could accommodate about 340 vehicles.

Truck weigh scales were put into service in 1981, primarily to weigh truckloads of sand, aggregate, and cement that were received on tract. These scales had the capability of weighing up to 60 tons gross weight.

A computer-controlled fuel dispensing facility was put into service in 1981. Diesel as well as gasoline storage tanks were connected to the system. Liquid petroleum gas (LPG) storage tanks were also on site to provide gas for heating the site buildings and facilities. Annual fuel consumption was 286,000 gallons of diesel, 126,000 gallons of gasoline, 177,000 gallons of LPG, and 400,000 million BTUs of natural gas.

A 9,000 gpd packaged Sewage Treatment Plant was required for construction. This plant was an activated sludge type plant, manufactured by Environmental Conditioners, Inc. Raw sewage was hauled by truck to the plant from holding tanks at the shaft sinking dry houses. A total of 60,000 gallons of sewage was treated and discharged annually.

A complete fire water system was installed to provide protection during construction as well as for the finished project.

Scheduled helicopter passenger service was provided between Logan Wash, Grand Junction and C-b. Emergency medical transportation was also provided.

Roadways were sprinkled with water on an as-needed basis (usually daily) during the summer months. Dust suppressant (Coherex/water mix) was applied on a scheduled basis or as conditions dictated. A road grader maintained the road surface, and loaders and trucks were used on an as-needed basis to clean ditches, culverts, etc. Snow removal and road-sanding crews were on 24-hour call for road maintenance during winter months.

In 1981, the Production Shaft, the Service Shaft, and the Ventilation/Escape Shaft reached design depth. Sinking employed conventional drill, blast, rock removal, and concrete lining techniques. Multiple small drill holes (under 2 inches diameter) were drilled in approximately 8-foot lengths, filled with dynamite, blasted, and mucked-out. This sequence was then repeated. After each successive 25 feet of shaft was sunk (30 feet in the Production Shaft), it was then lined with concrete. To my mind, it was basically digging a tunnel, but vertically rather than the normal horizontal.

Sinking in the 29-foot diameter Production Shaft progressed to the final depth of 1,867 feet (elevation 4,962 above sea level) by the end of September 1981. The 34-foot diameter Service Shaft reached a design depth of 1,765 feet (elevation 5,064) below the collar in May.

Shaft sinking for the 15-foot diameter Ventilation/Escape Shaft attained the design depth of 1,617 feet (elevation 5,088) in July, 1981. This included station excavation and concrete placement at the Upper, Intermediate, and Lower Void Levels. On September 2, 1981, shaft flooding commenced as a temporary measure to reduce discharge requirements from the holding ponds to Piceance Creek. The shaft area was secured and by year's end the water elevation was 6,297 feet above sea level, 426 feet below the collar.

Stations on all levels were completed to interconnect the Production and Service Shafts. The majority of the drifting (tunneling) was 20-feet high and 30-feet wide with smaller drifts for sumps and water collection areas. These stations were as follows:

Collar Level (6,829-feet elevation above sea level)
Midshaft Level (6,095-feet elevation)
Ignition Level (5,647—feet elevation)
Upper Void Level (5,481-feet elevation)
Intermediate Void Level (5,345-feet elevation)
Lower Void Level (5,208-feet elevation)
Bottom Level (4,966-feet elevation) (Production Shaft drift only)

Dewatering of the large shafts was, in brief, accomplished by utilization of 58 hp Flygt pumps to pump water from the sump at the bottom of the Production Shaft up to the sump at Lower Void Level of the Service Shaft. Then two Ash C-S pumps of 200 HP each pumped the water up and over to the sump at the Upper Void Level of the Production Shaft. From there 2 C-S's pumped up to the Midshaft Station at the Midshaft Level. From there, 4 C-S's and 2 C-65's (125 hp each) pumped the water over to the Service Shaft, up to the collar level and over to the lower mine water treatment ponds.

The shafts were classified as gassy by MSHA on January 2, 1980. The ventilation system was designed to comply with gassy mine regulations.

The Service Shaft was equipped with a 75 hp blower fan plus a propane air intake heater at the surface. At the Intermediate Void Level on the Service Shaft a 100 hp suction fan connected to a 36-inch ventilation tube moved the air from the bottom of the Service Shaft and exhausted it up the Production Shaft. Also at the Intermediate Level in the Production Shaft a 100 hp suction fan connected to a 36-inch ventilation tube moved air from the bottom of the Production Shaft and exhausted it up that shaft.

As the V/E Shaft filled with water, the shaft collar was sealed with foam. A high methane reading of 7.9% was recorded and the shaft area was declared "off limits". It was then decided that down-shaft ventilation would be reestablished. The foam seal around the collar was removed and 50 hp fans were put into operation and methane readings were reduced to approximately zero.

The number of acres disturbed during 1981 was 24 bringing the total to date to 188 acres (less than 4% of the Tract area). The areas disturbed in 1981 were associated with the fill material area used for temporary facilities (12 acres), additional topsoil storage area (5 acres), reinjection well drill pads (5 acres), and corehole drill pads (1 acre).

Other areas of construction activity during 1981 were included in previously permitted and disturbed areas. These included borrow areas used for material to raise the elevation in the Mine Support Area (included in the Mine Support and Ancillary areas graded in 1978).

Reclamation of disturbed land including storage of mined ore and overburden, and revegetation were concerns of the project. New areas reclaimed included the additional topsoil stockpile west of the Mine Support Area (3 acres) and the drill pads of two reinjection wells and three coreholes (4 acres).

Mined raw shale was backfilled into the storage pile in Little Gardenhire Gulch. The pile increased in area by one acre and approximately 30 feet in height in 1981. The total amount of raw shale that was backfilled during 1981, was 39,670 cubic yards (95,208 tons).

The total area of the storage pile in 1981 was 12 acres, containing 56,500 cubic yards of overburden and 126,221 cubic yards of raw shale.

The total acreage in graded condition in 1981 was 136 acres. Graded conditions consisted of areas that had been disturbed and then stabilized through the use of at least 50% rock in the soil surface and then mechanically stabilized through the use of dust pallatives and water.

Total acreage revegetated through 1981 was 34 acres, including such sites as drill pads and top soil stockpile areas. Revegetation processes included seeding with both temporary and permanent seed mixtures. Fertilizer was applied when appropriate. All revegetated areas were continually evaluated for production, i.e.: percent cover; number of species per square meter; mean production in grams per square meter, etc.

Air quality was affected by construction activities. It was the responsibility of the C-b Project team to monitor air quality and provide mitigation where appropriate. Activities that potentially affected air quality included sinking of the shafts, truck transport along haul roads, operation of the cement batch plant, operation of the feeder breaker, construction in the Mine Support Area, operation of the temporary power generators, and permitted open burning.

Air pollution permits required the use of control equipment and operating procedures. C-b held a continually updated Prevention of Significant Deterioration (PSD) permit from the EPA for the Ancillary Phase of MIS operations (defined in 1977 as up to nominally 5000 barrels/day oil production). Baghouses on the cement batch plant and emissions from power generators were typical point sources for which controls were required.

C-b obtained a Fugitive Dust Permit from the Colorado Air Pollution Control Division in 1977, revised in 1980. Pursuant to this permit, C-b paved the major access road to the tract. This work was completed in August, 1978. PSD and Fugitive Dust Permits required dust control on haul roads by regular applications of water and dust palliatives. The PSD Permit required quarterly reports to the EPA regarding both total water used and the amount and type of dust palliative applied. Water and dust palliatives were applied to the haul roads on an as-needed basis; dust palliatives were applied during 1980 and 1981. A busing system was instituted as one control measure to further reduce road dust by reducing vehicular traffic. Approximately 60% of the Tract personnel utilized the buses over time.

A typical situation arose in 1981 involving the use of a feeder-breaker to crush oil shale rock to minus 8-inch size. A permit was issued by the State of Colorado, which outlined acceptable operating parameters. Maximum throughput was limited to less then 1,000 tons per hour and annual throughput to 70,000 tons. Water spray bars were utilized as the approved emission control devices. No permit was necessary from the EPA since the annual emission level did not exceed the minimum permitted level of 25 tons of dust per year based on an emission factor of 0.1 lbs. of dust per one ton of rock. In another similar instance, a permit was issued by the State for open burning disposal of up to 10 cases of damaged dynamite in a manner acceptable to the Mine Safety and Health Administration.

Although no specific visibility-related regulations were issued by the EPA, visibility monitoring was conducted from the beginning of the project. No degradation in visual range was noted. Considerable concern was expressed throughout the project life that oil shale facilities would negatively impact visual attributes of western national parks and monuments. Existing wind-blown pollution from urban areas of California and power plants in the Four Corners area were cited as justification for the concerns. In 2001, Californians paid for these concerns through rolling blackouts and other power scarcity.

Compliance with the NPDES permit was continuously monitored and results reported to regulators. Close contact with the State Water Quality Control Division was always maintained regarding progress on NPDES permit processing and in-stream classification of Piceance Creek Requirements for water augmentation were formulated, reviewed and redesigned continually. Planning on water storage projects proceeded along several lines, including

1)  Yellow Jacket Study: this study was funded by the State of Colorado and restricted to selection of storage sites on the White River or its tributaries above the confluence with Piceance Creek.

2)  White River Study: this plan was a joint effort between industry and the State's Yellow Jacket plan. Work included evaluation of alternatives and delivery systems to cover Piceance Basin oil shale projects.

A Spill Prevention Control and Countermeasure Plan included a description of the potential for accidental spills or release of oil and other hazardous materials as a result of Tract development and associated off-Tract pipelines and terminals. This plan summarized the potential source of accidental spills, reviewed current regulations and standards that would apply to project activities, defined and inventoried the hazardous materials within the plant, and presented the project spill prevention, control, and contingency plans for the plant and associated pipelines.

In the event of an accidental spill of oil or hazardous material in quantities greater than those specified by the regulations, various governmental entities were required to be notified. Spills consisting solely of oil were reportable if they reached or had the potential of reaching a waterway in quantities, which cause a film, sheen or discoloration of the water. Spills involving hazardous materials were reportable if they occurred on the land or could reach a waterway in quantities exceeding those specified by the regulations.

| Notification | Spills-Situation |
|---|---|
| National Response Center (NRC) | 'Reportable Spills' |
| Regional Response Center | When the NRC cannot be contacted |
| Colorado Department of Health | 'Reportable Spills' |
| Colorado Division of Wildlife | Danger to fish, etc., in surface water supplies |
| Water Quality Control Division, Colorado Department of Health | Contamination of water supplies |
| Colorado Highway Department | Move vehicles, control traffic |
| Oil Shale Supervisor | All spills |
| BLM, US Forest Service | Certain Cases |
| Local city fire, police, and health departments | Major spills |

The plan required that all spills be responded to by an in-plant spill response team specially organized and trained for this purpose. A Spill Response Coordinator (SRC) was delegated the primary responsibility for deciding the action required and assembling the necessary team elements. The team members included the following C-b staff members:

> Spill Response Coordinator
> Cleanup Coordinator
> Government Liaison Coordinator
> Public Relations Coordinator
> Legal Coordinator
> Environmental Protection Coordinator
> Procurement and Logistics Coordinator
> Document Coordinator
> Accounting Coordinator
> Training Coordinator
> Safety and Security Coordinator

In addition, all appropriate governmental officials listed above would be involved in a spill. I always gratuitously suggested that they have a vacant theater on contract in which meetings of all the concerned parties could be held.

Solid waste (trash) accumulated in waste bins was trucked off-site as frequently as necessary to approved land fills in Rifle.

As part of the project's commitment to protect historic, scientific and aesthetic values, numerous studies were undertaken on the Tract and surrounding area and reported.

Health, safety and security were given maximum attention by all C-b employees. All contractors conducted regular safety meetings for their employees with the active participation of the C-b Safety Department. All new employees were required to receive health and safety training prior to being assigned work duties.

In 1981, C-b had a Health/Safety/Security Supervisor, two safety inspectors, a Security Supervisor, ten Security Guards, and one secretary. Two major contractors also had full-time safety staffs.

The C-b Health Department consisted of a Grand Junction based Industrial Hygienist and an Industrial Health Coordinator who spent 25% of her time at the C-b Tract.

Emergency medical service was provided twenty-four hours a day by one paramedic and eleven emergency medical technicians (EMT's). A fully equipped ambulance was available for off-site and C-b Tract emergency medical treatment. An ongoing EMT training program was established with the assistance of Occidental physician advisors for the emergency medical personnel on the C-b jobsite. Part of the EMT/ paramedic training program consisted of monthly training sessions with demonstrations and shaft extrication classes.

An EMT coordinator monitored the First Aid Trailer, ambulance, all medical supplies and equipment and assisted with the EMT training.

St. Mary's Air Helicopter located at St Mary's Hospital in Grand Junction was available for medical emergencies twenty-four hours a day. C-b operated a Long Ranger helicopter available for backup emergency medical transportation.

Security, Safety and Industrial Hygiene Manuals were prepared and utilized by all staff. Industrial hygiene research at Occidental's Logan Wash oil shale facility was fully utilized to provide input to properly design and implement an industrial hygiene program for the C-b project.

A Fish and Wildlife Protection Plan was developed to provide procedures to avoid or minimize adverse effects on fish and wildlife caused by the development and operation of oil shale facilities on Tract C-b. The habitat management plan used the baseline environmental data as a frame of reference. It delineated habitat losses that could occur and mitigation efforts possible to either replace in-kind or improve alternative habitat for selected species of animals. It was hypothesized that the main C-b access road could impede deer movement through the pinyon-juniper vegetation type north of the site. Also, it was felt that a major ecosystem impact could result from deer/vehicular collisions. These impacts did not materialize. Studies showed that the deer did not significantly alter their use patterns near the access road. Various mitigation activities were undertaken throughout the life of the project. Selected areas were intensively studied over time including sampling for plant productivity, species composition, and deer and lagomorph abundance.

A land application (sprinkler) system was installed and operated during summer months. The main objective of the irrigation system was to provide a method of disposing of excess mining water until such time as the water could be used for retort processing or for reinjection. The sprinkler system proved to be a good mitigation project because the vegetation productivity increased due to the additional moisture. Livestock utilization also increased in the sprinkler area. Permanent sprinkler area transects were sampled for deer and lagomorph pellet group densities, productivity and utilization, small mammal and avifauna abundance.

C-b, in cooperation with various agencies including the Colorado Division of Wildlife and Oil Shale Office plus the Rio Blanco Oil Shale Company, tested a new type of reflector along the Piceance Creek Highway. These post-mounted reflectors were successful in Austria, but it was believed that a local controlled study was needed to fully evaluate their effectiveness. This type of reflector refracted red light away from the road which seems to last for longer duration than the single flash from a conventional mirror reflector. The reflectors were installed along both sides of the highway to light up progressively as a vehicle proceeded down the highway creating an optical warning fence at night. The study design included installing four one-mile sections of these reflectors (both sides of the road) in areas having the highest record of roadkills. These mile sections were randomly covered (jacketed) for a week, with two sections in operation at one time. to reduce bias due to weather. Results were statistically non-conclusive.

Local springs and seeps were continually evaluated to determine their potential use for livestock and wildlife watering.

Mitigation projects were always under consideration including: approved burnings, planting seedlings in chained areas (much of the 5000 acre tract had been previously "improved" by dragging a big logging chain across the land with large tractors to ostensibly enhance grazing), fencing for better cattle distribution, additional stock tanks and water wells, proposed dams for water storage which could create waterfowl wetlands and additional fishery habitat. One of my favorites was the clever use of discarded large equipment rubber tires for livestock and deer watering tanks. The tires were cut in two and placed strategically throughout the tract. In addition to providing necessary water tanks, the scheme recycled otherwise unusable scrap tires.

With admirable foresight an Abandonment Plan was prepared in the Development Plan Modifications submitted July, 1977.

As the leasing monitoring system developed, full attention to impacts on the surrounding public community population as well as the impacts of any development on the ecosystem was mandatory.

Socioeconomic activities undertaken by the C-b staff and subcontractors included analysis of the workforce and the associated population buildup, government Mitigation Task Force support, workers programs, community donations and public relations.

To provide some prospective about the task we were contemplating, the following is one of many employment predictions made by and for us. It shows the level of impact we were planning for the local area. In addition to the direct employment figures represented by these numbers, it was understood that in addition to those workers who migrate into the area for construction and operations jobs, there will be a significant number of people filling jobs resulting from the overall economic growth and diversification created by the development of our project. This induced workforce was often represented as about one-fourth of the total direct workforce. In other words, the overall impact of the C-b project on local communities in 1987 according to the following table would be 1.25 times 1800 or 2250 workers plus their families. It was generally assumed that construction workers would leave when their work finished, and the others would remain. So the problem was always two-fold; provide for a large temporary force and provide permanent infrastructure for a nearly as large permanent population.

| Year | Construction | Operations | Total |
|------|--------------|------------|-------|
| 1984 | 250 | 250 | 500 |
| 1985 | 900 | 600 | 1500 |
| 1986 | 1900 | 600 | 2500 |
| 1987 | 1000 | 800 | 1800 |
| 1988 | 200 | 1000 | 1200 |
| 1989 | 0 | 1000 | 1000 |

During early C-b construction days, the majority of site employees resided in Rifle, but the percentage of total employees residing in Rifle declined as time went on.

The C-b employee bus system in 1981 consisted of ten operating buses. An estimated total of $775,000 was expended that year on lease and operating costs for the bus system. Based on the transportation data, approximately 60% of site personnel utilized the buses. C-b participated during 1981, along with four other oil shale projects, in joint financial sponsorship of the construction of a highway by-pass route in Rifle.

C-b leased 111 apartment units in Rifle and Meeker and operated a 103 unit King's Crown mobile home park in Rifle for employee housing. The total net cost of operating employee housing for a typical construction year was $250,000. Additional housing programs were developed as the project proceeded.

It was necessary to provide in-house Community Relations staff. A manager was hired and stationed in Rifle. Part of her responsibility was to provide ongoing technical support to the impact mitigation task forces in Rio Blanco and Garfield Counties.

Rio Blanco County provided a resolution to issue pollution control bond financing for C-b. As inducement for the financing, an agreement was executed between C-b and the county calling for the development of a socioeconomic impact mitigation program. This program was developed jointly by representatives of C-b and Rio Blanco County.

Public interest in the project was very high. During a typical year, 1981, our staff conducted 218 tours at the tract. Of these 41% were industry related, 30% governmental, 27% educational and 2% media related. There were a total of 1,852 visitors involved with these tours.

In addition, various members of the staff made a total of 168 lectures and/or presentations. In addition, there were five expositions at which models of the tract were displayed. A total of 135 photographic work requests were completed and a chronological historical file of 35 mm color slides was updated monthly to record progress. Several film crews, including all three national networks and Armand Hammer Productions, were given assistance while filming on-tract.

All activity was conducted with strict adherence to environmental, permit, and lease regulations. Environmental impacts, where they existed, were strictly confined to the immediate tract and within limits defined in the Detailed Development Plan. Environmental monitoring played a key part in the project throughout the total life of the project. As noted previously, from November 1, 1976 through March 31, 1978, the C-b Tract was under a period of suspension of the Federal Oil Shale Lease. The monitoring conducted during this period was executed under a program known as the Interim Monitoring Phase. The Development Monitoring Program for Oil Shale Tract C-b was submitted to the Oil Shale Supervisor on February 23, 1979 and approved by the Oil Shale Supervisor on April 13, 1979 subject to thirteen Conditions of Approval contained in the approval letter. Semiannual environmental data reports were submitted twice yearly.

All air, water and microclimate data were stored in a computerized database called RAMIS. Biological data were partially in manual databases. QA audits for data collection were regularly conducted.

In the Environmental Baseline Monitoring Program emphasis was placed on key indicators of environmental quality and/or change. The objectives of environmental monitoring were defined to provide: (1) a record of changes from conditions existing prior to development operations, as established by the collection of baseline data, (2) a continuing check on compliance with the provisions of the Lease and all applicable Federal, State and local environmental protection and pollution control requirements, (3) timely notice of detrimental effects and conditions requiring correction, and (4) factual basis for revision or amendment of requirements.

The approach taken in the Development Monitoring Program carried an oxymoronic title of a "simple, multicomponent, conceptual model." Inputs consisted of the environmental database, the Lease Environmental Stipulations, the details of Tract operation, and applicable local, state, and federal regulations. The outputs or actions constituted the Development Monitoring Plan and its implementation (findings) as a result of monitoring. A decision matrix consisted of the three major criteria to which candidate variables for monitoring were subjected (relatability, observability, and measurability). A significant feature of this conceptual model was its feedback capability. That is, variable levels were assessed against expected levels. In the event that out of scope levels were obtained, a systems dependent mode of either more intensive monitoring, use of additional stations, or added variables (or all three) was triggered. Feedback from input data was used to obtain improved inputs and provide continual review and refinement of the monitoring programs as additional information was collected and analyzed.

The Development Monitoring Program was believed to have been brought into sharper focus with the identification of Class 1 indicator variables. These were key environmental variables collected at representative stations in at least monthly sampling frequency. Time series plots, generated by the computer from the database to a common time scale, and updated in the semiannual data reports to provide visual analyses of trends and interrelationships. As a statistical screening process, linear short- and long-term trends were examined at a five percent level of significance for air and water and to 20 percent for biology.

A photographic record of Tract changes was maintained. A 360-degree horizontal pan was photographed in color on a yearly basis at 35 photo points. Color infrared panoramic photographs of the vegetation around springs and seeps were obtained three times during the growing season. Landsat digital imagery was used to monitor vegetative condition in the Tract vicinity.

Water quantity and quality data were also monitored for the purpose of impact evaluation. Streams, springs, seeps, alluvial and bedrock aquifers, shafts and impoundments were monitored. Baseline studies indicated the mean flow for the reach of Piceance Creek adjacent to the Tract to be approximately 13 cubic feet per second. Continued monitoring indicated no significant change in mean annual flows.

The flows of groundwaters are governed by the stratigraphy of the Tract. Insights into groundwater hydrology as functions of depth and stratigraphic zone were gained continually as the three shafts were sunk into the shale deposit. Data collected strongly suggested that the tight confining zones of oil shale highly restrict the vertical movement of groundwater. The most significant data in this regard came from two well pairs along Piceance Creek north of the Tract. Each well pair consisted of a deep bedrock well and an alluvial well within close proximity of one another, and was designed to determine whether the dewatering activities in the deep bedrock aquifers were affecting the alluvial aquifers. For both springs and alluvial wells, no significant long-term trends in water quality values were found. Major constituents examined for trends were temperature, pH, conductivity, Dissolved Oxygen Content, arsenic, fluoride, boron, Total Dissolved Solids, molybdenum, sodium, sulfate, and ammonia.

Aquatic systems were examined regularly. A multitude of factors such as irrigation, cattle grazing, springs, and Tract C-b water discharge could affect Piceance Creek aquatic systems.

Air quality variables were continuously monitored, examined and evaluated. Compliance with Federal and State air quality standards was maintained on the C-b Tract. Meteorological data were collected for the life of the project as well as noise data.

Flora and fauna were exhaustively measured year around. Population densities and inter-species relationships were analyzed, for example:

- Browse utilization was not significantly affected by development activities,
- Migrational distribution of mule deer was not significantly affected by C-b.
- Deer herd size and weather appear to exert more influence on road kill than vehicular traffic.

Wintering deer in the tract area appeared to be using the habitat much as they did during the pre-development period. Coyotes and lagomorphs were the two medium-sized mammals monitored on Tract. Coyotes are of ecological significance because they are a major predator on Tract C-b. No significant changes in abundance of lagomorphs were detected. Small mammals trapped included deer mice and least chipmunks. Development activities did not cause any significant changes in songbird diversity or density. Mourning dove population continued to fluctuate yearly without any definable patterns. Development activities seemed to have little effect on raptor activity in the area.

There were no major changes in the herb layer species composition over at least seven years. There was a trend showing a decrease in total plant cover. This may have been related to changes in cover estimation or, to the successional dynamics of the chained rangelands. Shrub species composition did not change, however total shrub cover and density increased. None of the noted changes in the vegetation appeared to be related to the development of the site.

No threatened or endangered species of plants or animals were observed on Tract. Revegetation success of the topsoil piles had been achieved regarding herbaceous cover, diversity and productivity.

As perhaps can be concluded from the exhaustive (and exhausting) dissertation above, the C-b project was not just designing a technically feasible commercial mining/petrochemical facility. We were fully complying with the terms of the lease, especially the prototype aspect which was designed to show that oil shale could be developed within acceptable limits on impacts to the human and ecological communities in the area. I will always be proud of that effort. I would like to have seen it completed so that these goals of the leasing program could have been met.

# CHAPTER 10

# Another Change in Management

DESPITE ALL THIS activity, it was undeniable that the generosity of the partners was due to unusual circumstances. The fact that the last two lease bonus payments to the government could be avoided if appropriate expenditures were made on the project was the first thing that made money more available than what otherwise might have been. When Tenneco agreed to put $100 million into the project as a buy-in, which was to be spent without matching Oxy funds, that money was made more readily available. So significant expenditures were made on the project just to keep it going and the resultant momentum made it easy for us project people. However, careful money managers in both the Oxy and Tenneco organizations were asking where and when the payback would be realized. Oxy didn't have lots of cash. They were in the process of joining up with Cities Services with the resultant power plays between upper managements with large amounts of uncertainty filtering down to those of us in the trenches. Tenneco apparently saw significant opportunities in synthetic fuels. They were putting big money into a coal gasification project in North Dakota. They also showed great determination in furthering our oil shale project. But, it seemed clear that they felt that the road to success was through the federal government. They apparently felt strongly that federal financial assistance was crucial to the success of these pioneer, high-risk projects.

At the same time, the estimated costs of the plant continued to grow. As mentioned previously, this was due to several things, including the double-digit inflation of the times, basic lack of understanding of the design of the plant, and management's tendency to change contractors as one might change shoes. The latter factor was the old "kill the messenger" story. If you didn't like the estimate message delivered by your contractor, fire him and hire another from the many anxiously waiting in the wings. We changed contractors far too many times. And each time cost time and money getting a new team up to speed and convincing them that most of the ideas they were suggesting had already been examined and discarded for cause. A typical example is the following excerpt from an unsolicited presentation from a major contractor who wasn't on the team at the time, but really wanted a piece of the action;

*"Technical Considerations Requiring Further Evaluation to Fix Specification for Basic Engineering Design Prior to Detailed Engineering*

To date, several conceptual design studies for the aboveground facilities of a commercial MIS shale oil project at Tract C-b have been executed. These studies have confirmed the basic economic and technical feasibility of the project. However, one must not lose sight of the *conceptual* nature of the studies. The studies did not extend beyond determining that the new technology *could* be resolved by application of commercially proven, existing equipment and engineering techniques.

The design of a processing facility, which has been accomplished many times, dictates equipment specification to be initiated during the early stages of the detailed engineering phase normally prior to the issuance of P & I (Process and Instrumentation) flow diagrams. This procedure allows a shorter schedule with little risk for changes in equipment specification, whereas changes cause recycle, wasted effort, and extra expenditures in time and money. The Cathedral Bluff (sic) plant will be a prototype facility for which it is considered advisable to fix Specification for Basic Engineering Design prior to detailed engineering. This concept allows the development of a firm project design basis so that equipment specifications can be initiated in the early stages of detailed engineering with the same confidence, as for a non-prototype facility, that changes will not be encountered. In effect, this is a *project management priority* and does not in any way detract from the conclusions resulting from the conceptual studies.

Many major and minor technical considerations became evident in the course of the conceptual studies. A few of the major items are listed here. They are divided into two groupings: (1) Those requiring intensive evaluation of operating data as well as engineering design study and (2) those requiring engineering design study alone.

1)    *Items Requiring Intensive Operating Data Evaluation*:

1.    *Basic Material Balance*

The conceptual studies were based on single rates, compositions, and conditions of MIS off-gas and mine water. At this point, the data needs (sic) intensive examination to establish ranges of rates, compositions, temperatures, pressures and the like, during the course of operations, to assure that changes occurring due to operation and variation in the shale strata will be accommodated by the facilities design. Specific design items involved are:

> Off-gas compressors
> Off-gas cleanup system (in particular to separate oil mist fog
>      if present)
> Sulfur removal system

System to remove heavy metals if present in the off-gas

Total water management system

Aboveground retorting facilities and their integration into
the total plant.

## 2.   *Equipment/Materials Application Considerations*

To assure successful startup and operation, it is wise to confirm that
potential fouling and corrosion concerns can be dealt with. In
addition, the need to break potentially stable oil-water emulsions
can be anticipated. Thus equipment to be applied should be
carefully analyzed and its successful operation confirmed. Items
for special consideration are:

Is fouling of the incinerator burners or the off-gas compressors
a potential problem?

What equipment/systems will be required to break the oil-
water emulsions?

Will the incinerator burners operate stably over the range of
anticipated off-gas composition? How much pilot fuel
will be required? Control considerations.

Are there any parts of the system where special corrosion
protection is advisable?

## (2)   *Items Requiring Engineering Design Study*

Items such as the following must be resolved as part of the
Specification for Basic Engineering Design. If not, serious negative
project schedule impact could result:

Detailed design and layout of the MIS off-gas ducting is
necessary to confirm basic design for expansion, support,
insulation, tracing, etc. Allowance for future retort additions
should be included. Interaction with compressor design and
design of shutdown/trip systems must be considered.

A comprehensive study of the control/shutdown/trip systems for the off-gas compressors and mine air supply compressors is necessary to ensure safe, reliable, trouble-free operation of the total system.

The total system involves extensive piping and equipment systems handling water, steam and wet gases. Extensive winterization to assure reliable, trouble-free operation is vital. One can anticipate that winterization can impact equipment and piping design, plot plan, utilities system design and the like. Critical areas should be examined for winterization so that the impacts can be included in the Specification for Basic Engineering Design.

### Summary

The present schedule allows for initial firing by October 1, 1984, based on starting detailed engineering in January, 1981. This schedule is based on the assumption that information is available to fix the basis for engineering design sufficiently to allow initial production of equipment specification at the start of detailed engineering with minimum risk for later changes. It is desirable for a prototype facility such as the proposed plant, to fix the Specification for Basic Engineering Design prior to start of detailed engineering to insure minimum schedule risk. Early 1980 start in preparation of the Specification for Basic Engineering Design is recommended to minimize schedule risk."

Clearly, this contractor (who incidentally got some work and C-b money) didn't think much of our project premises and bases of design. The contractor was right. The year called for to develop basic engineering design was hardly enough. *There were no successful oil shale processes in existence from which anyone could extract basic design parameters and establish a firm design.*

In addition, there was a major difference of opinion between Oxy and Tenneco management regarding the capabilities of contractors. Dr. Hammer always felt that the Bechtel Corporation, and especially their senior management gave him the best deal. Tenneco favored others.

All of these factors plus others, lead Tenneco to put on a power play. From my balcony seat, I'd have to say that the doctor blinked, because Tenneco ended up as operator of Cathedral Bluffs and we Oxy folks were either demoted or fired. I ended up assigned to development of a *new* project scope (the latest in a long line) and consequent estimate. Although I lost my corner office to the Tenneco lawyer, I was lucky. It was a discouraging time because of the nonsense between the partners. But a couple of good guys, Fritz Kloer and Bob Bloom, and I moved across the street into the old Exxon offices and set up our own project office. We filled the office at times with contractor people who turned out work scopes and estimates on demand.

# CHAPTER 11

# Federal Assistance

A MAJOR FACTOR in our project and all the other oil shale projects became some sort of involvement with the federal government. Notwithstanding the fact that C-b was being built on federal ground under the terms of a unique leasing agreement which placed many extraordinary, first-of-a-kind economic burdens on the project, federal subsidization was not part of the agreement, except for the clause which allowed us to offset the last two lease bonus payments by building appropriate facilities on the site. The lack of direct federal support was not a surprise, but was instead one of the foggy things that entered into the original bidding on the prototype leases. I wasn't there and I can't relate to the reasoning that went into the very high bids. My Monday morning quarterbacking is as good as anyone's, and I have been wondering for nearly thirty years now why things went as they did. Don't get me wrong, I'm certainly glad things went as they did in the early 70's. I wouldn't have missed the experience I was privileged to enjoy for anything.

The federal government responded. Several agencies were established, not the least of which was the Department of Energy, and administrative poster campaigns such as Project Independence which was to solve energy supply problems in the same manner as the Manhattan Project built the atomic bomb and the Apollo Project put men on the moon. President Nixon had gone on national television in 1973 to plead, "Let us set as our national goal, in the spirit of Apollo, with the determination of the Manhattan Project, that by the end of this decade we will have developed the potential to meet our energy needs without depending upon any foreign energy sources. Let us pledge that by 1980, under Project Independence, we shall be able to meet America's energy needs from America's own energy sources." Project Independence was based on (1) Significant expansion in the nuclear power industry, (2) Development of oil shale, (3) Increases in the production and use of coal, and (4) Greatly increased domestic oil and gas exploration with concomitant production.

Shell Oil, as early as 1976 began talking about the need for federal government involvement. My boss at the time, Bob Meeker, in an interview for the "Shale Country magazine,[viii] said;

> "The shale-oil business is quite different. It is a *development*-not yet an on-going cash-producing business. A shale-oil plant represents a tremendous investment—$1 billion per plant (at least $2 billion five years later)-and it's a long-term investment, which under the best conditions, will not pay off for many years. Managing that kind of investment requires different strategies. After all, $1 billion is a staggering amount of money. There are very few entities in the world that can appropriate that much capital. Even if stockholders' funds and retained earnings were available in that amount, as a manager, I would be violating their trust if I unduly put their funds in jeopardy by investing in an oil-shale project that could not make an adequate return on their investment. It's also important to realize that profits are *essential* to the workings of our private enterprise system. For example, the billions of dollars needed for new American energy projects can only come from profits and

from lenders, who will only give money to a company that is economically sound-profitable . . . ."

The article goes to on to point out that private capital will not now supply funds for a commercial oil-shale project. In addition, the costs of plant construction were escalating rapidly, but there was not a proportionate increase in oil prices. Citing a Tosco report, Meeker points out that;

"Other factors that act together to preclude private investment in oil-shale are the risk of loss inherent in any pioneer project and the high price tag on oil-shale plants. Even if oil prices were now high enough to justify projection of a 15 percent discounted cash flow return rate (on investment), there is no assurance investors could be found who are willing to risk $1 billion in a pioneer project."

The article concludes with editorial observations;

"that the federal government will have to recognize that it is in the national interest to develop synthetics and this means that government financial incentives are critically needed. Indeed, they are absolutely necessary *if* commercial development of oil shale is to occur and meet America's long term energy needs, and *if* this development is to occur in an orderly manner, with a minimum waste of money and minimum impact on the environment and local communities."

I know this all seems confusing. Did Shell want federal incentives or not? If they did, why did they bid over a $100 million as a pure bonus on a project where such government support was clearly not planned? It partially explains why Shell elected to withdraw from both the Colony project and C-b before they got too far along.

The federal government was not ignoring the situation. As noted, an Energy Research & Development Administration had been formed by Congress. This agency began developing a position on synthetic

fuels, including oil shale, as described in the following article published
in the April 1976 issue of "Shale County".

"In February, the American Chemical Society held a program
in Colorado Springs to discuss the federal Synthetic Fuels
Commercialization Program. At that time, and at press time, the
U.S. Congress was once again considering funding legislation for
this program. The Senate bill did not mention oil shale; the House
bill did, but endorsement of the shale section was wavering.

Congressional and Administrative support for oil shale is
considered essential by the industry—not just to insure
commercialization but, in fact, to insure survival. However, as noted
before, such support is a sometime thing; thus oil shale's permanent
position in the commercialization program is unclear. Nevertheless,
it is important for SHALE COUNTRY readers to understand the
rationale behind the program, including its oil-shale portion.
Therefore, this article will summarize key points of the program
and key statements from those responsible at the Energy Research
& Development Administration (ERDA) for its implementation.

**What and why is ERDA?** The U.S. Energy Research &
Development Administration was established by legislation in 1974
as the lead federal agency for energy research and development. Its
mission is to develop all energy sources, to make the nation basically
self-sufficient in energy, and to protect public health and welfare
and the environment. ERDA programs are divided into six major
categories: energy conservation; fossil energy; solar, geothermal and
advanced energy systems; environment and safety; nuclear energy;
and national security. ERDA programs are carried out by contract
and in cooperation with industry, universities and other
government agencies.

**What is the Synthetic Fuels Commercialization Program?** It is an
incentive program to encourage the private sector to construct and
operate first-of-a-kind commercial-scale synthetic fuels plants. These

plants include shale conversion to synthetic petroleum; coal conversion to gas or oil; solid waste conversion to gas or oil.

**What are the major objectives of the program?**
—To gain information and to initiate development of industry infrastructure by investigating environmental, economic, institutional, technical and other potential problems, and by promoting private sector synfuels experience.
—To supplement existing and planned domestic energy production by reducing reliance on imports and by providing less expensive energy if the world price rises.
—To improve the U.S. international position in energy matters by demonstrating U.S. capability to tap its vast resources and by establishing U.S. leadership among consuming nations.

**How is this program different from other ERDA programs?** As contrasted with ERDA's energy R&D program, which is aimed at developing new or improved technologies in order to demonstrate *technical* feasibility (such as ERDA's oil-shale in-situ research), this program would create through federal incentives, *commercial demonstration* of a limited amount of synthetic fuel production using already-developed technology that can be applied between now and 1985.

**How did this program begin?** In his 1975 State-of-the-Union message, the President called for accelerated development of U.S. energy technology and resources as well as strong energy conservation measures. As part of the plan to accomplish this energy self-sufficiency goal, the President proposed a Synthetic Fuels Commercialization Program. Thus, an Interagency Task Force, composed of 50 members from 13 federal agencies, was formed in February 1975, under the direction of the Energy Resources Council, to examine alternatives for implementing the President's synfuels objectives. The final results of the Task Force's efforts were released as a *four-volume* "Synthetic Fuels Commercialization Program" report in January 1976.

**What is the report's key recommendation?** The Task Force stresses that federal financial incentives are needed *now* to initiate a limited commercial demonstration program to provide the basis for a synthetic fuels industry. Thus, the Task Force has outlined a federal incentive program designed to insure initial production of the equivalent of 350,000 barrels/day of synthetic fuels before 1985. This is the first phase of a possible two-phase program that could ultimately encourage up to 1 million barrels of synfuels by 1985.

**Why are synfuels needed?** "America's dependence on foreign sources of oil and gas continues to grow. Domestic production of oil and natural gas has been declining since the early 1970s. By 1980, *if* past trends persist, the U.S. will depend even more on foreign energy sources than today. Although the development of Alaskan and Outer Continental Shelf oil and gas, improved energy conservation, expansion of nuclear power facilities, and greater direct burning of coal can buy time, our dependence on foreign sources of oil and gas will continue to grow if synthetic fuels are not produced in substantial quantities," says Dr. William T McCormick, Jr., who chaired the Task Force while at the Office of Management and Budget. He is now at ERDA heading up its newly-formed Office of Commercialization.

**Why are federal incentives needed?** According to the report, at the present time there are a number of serious impediments to private sector commercialization of synthetic fuels. The uncertainty of future prices of world oil is perhaps the most important factor discouraging private investment. If world oil prices were to fall substantially, large plant investments could not be paid off from revenues of low-price, but high-cost, synthetic fuels. In addition to the financial risk, there are numerous environmental, regulatory, and labor and materials uncertainties. Federal government involvement is needed to overcome these.

**What type of federal incentives are needed?** The Task Force examined a variety of options, including tax changes (investment tax credits, construction expensing, and accelerated depreciation) and government-owned management structures. The recommended incentive for shale oil and syncrude was for the federal government to provide up to 50-percent non-recourse loan guarantees and price guarantees, competitively bid. The major features of this incentive are:

- It protects the lender from some capital risk through the non-recourse loan guarantee.
- It protects the corporation in the market place through price guarantees by assuring a fixed price.
- And the procedure of competitive bid for a price guaranty should result in lower shale oil and syncrude prices, which will reduce the cost to the government of the price support program.

From the government point of view, the major strength of the recommended incentive is that it:

- Encourages industry competition and broad participation through its loan-guarantee provision for those firms needing risk sharing.
- Reduces or eliminates government costs as market prices approach the production prices of syncrude.
- Does not require government management or operation of plants, thus minimizes federal involvement.
- Provides an anticipated subsidy limited to the production-plant life and thus would not result in permanent subsidy to industry.

**What does the report say about the environmental impact of such a program?** The environmental impact statement provides a comprehensive description of the potential impacts—from landscape alterations to health-care demands—that may result from a synthetic fuels program. In turn, the Task Force has recommended a detailed environmental protection strategy, which includes such steps as site-specific EIS's, an environmental advisory board, carefully prepared development plans for individual plants, coordination with local and state agencies, as well as comprehensive environmental monitoring and surveillance. A major goal of the commercialization program, in fact, is to yield environmental and other data that will allow development of the synthetic fuels industry as the program's experience in the nation's interest warrants it.

**What is the states' role in the program?** The Task Force received comments from the states through the National Governors Conference and the western governors, as well as other organizations, and made significant changes in program recommendations based on those comments. As a result of the states' interest in a synthetic fuels program, the Energy Resources Council has also established a new intergovernmental coordinating committee to receive state input on synfuels and other energy issues that require federal/state interaction.

**What is the future of synfuels commercialization?** ERDA says, "Despite the deletion of the synthetic fuels commercialization program by the House in December 1975, the Administration still strongly supports a substantial loan-guarantee program as a major step toward this nation's energy self-sufficiency. ERDA expects congressional reconsideration of synfuels legislation this year and strongly supports its early enactment." Time, of course, will tell the amount of commitment government has to synfuels and to shale. As Bill McCormick said at the conference, "if the Synthetic Fuels Commercialization Program is defeated again, it may be shelved—for years."[ix]

In a sidebar attached to this article, the governmental representative spoke some strong words, the likes of which continued to encourage industry representatives. Unfortunately, government was not able to meet all their promises.

> **"Baloney to other energy options,"** says Bill McCormick, head of ERDA's Synthetic Fuels Commercialization Program. 'If you look carefully at the energy situation in the next 25—30 years, it is very clear that alternative energy sources such as nuclear and solar will not be sufficiently developed to handle the supply-demand cap we are facing. Thus, we are convinced there will be a *significant* demand for synfuels in the 1990s, a demand that will require 100 major synfuels plants. And we are convinced we've got to get the industry going so we can start to address synfuels legal, environmental and social issues. If it doesn't happen soon, there will be a crisis later. Then the government will have to come in and develop synfuels alone and that would be a tragedy for our free-enterprise system. We must start to develop an industry-government partnership *now* and get a limited number of plants underway. Then we will have the information we will need to make intelligent decisions in the 1980s about full-scale synfuels development."

It was and still is simply amazing to folks like me that a senior honcho in the Ford administration was telling us just what we wanted to hear. Is it any wonder that we rushed out and committed large amounts of dollars to try to meet the stated need for "100 major synfuels plants" by "the 1990s". It looked like Mr. Meeker's prayers were being answered.

The federal government continued to try to support alternative fuel development in their remarkable way. However, in 1980, the government's efforts centered on the Synthetic Fuels Corporation. This federal bureaucracy tried in its way to encourage alternative fuel development as described in the Summer 1983 issue of "Shale Country" magazine.[x]

> "When the U.S. Synthetic Fuels Corp. was created under the Energy Security Act of 1980, its days were already numbered. The corporation, which is a quasi-governmental agency, was formed to award loan and price guarantees, make purchase commitments,

direct loans, and participate in joint ventures with private synthetic fuels companies.

The money is to be used to foster development of a diverse set of technologies and resources and to meet national production goals of 500,000 barrels/day by 1987 and 2 million barrels/day by 1992. After providing this assistance, the corporation is scheduled to close its doors no later than September 30, 1997, leaving behind an independent and commercially viable synthetic fuels industry.

In order to allow Congress to review and evaluate the corporation's success in meeting these objectives, the limited lifespan of the Synfuels Corp. is divided into two phases. During the first phase, which runs from 1980 to 1984, Congress has appropriated $15 billion for the corporation's activities. Then, in 1984, the corporation must submit a strategy to Congress for achieving the national synfuels production goals and explain what, if any, further funding should be provided for the remainder of its tenure.

If approved, Congress can appropriate up to $68 billion more so that the corporation can continue its financial programs through September 1992. After that, no additional funds may be committed by the corporation. The staff will only monitor projects that received funding and wrap up related activities by 1997.

## A rough take-off

While the corporation's mission sounds straightforward, some curves have developed in its path to progress. Although the Synfuels Corp. sent out a request for proposals for financial assistance to qualified applicants in 1980, it just began giving out awards this year.(1983)

Part of the hold-up came with the change from the Carter to the Reagan Administration. The Synfuels Corp., formed under Carter, had barely gotten off the ground when Reagan was elected in 1980. The board of directors resigned, and a new board then had to be appointed and confirmed by the Senate. Since the first general solicitation (request for proposals) closed in March 1981, no action could be taken on applications until the new board convened for the first time that fall.

The new board now has the ball rolling and has since sent out two more general solicitations, but the process is painstaking by nature. It begins when the Synfuels Corp. issues an invitation to submit proposals for financial assistance. Interested project sponsors submit a proposal, which, says Karen Hutchison, director of media relations for the Synfuels Corp., describes the present status of a project, the technology being used and related financial information.

In two separate reviews, proposals are evaluated according to stringent criteria. But first the corporation screens all applications to insure they meet certain eligibility requirements. For example, a project must involve transformation of domestic coal (lignite and peat are included), oil shale, tar sands or heavy oil into a synthetic fuel product.

If the project qualifies, the first review begins, and the project is evaluated on the basis of maturity. To determine if the project can meet the corporation's maturity criteria, the staff looks at the amount of equity sponsorship, which is the amount of money the sponsor is committing to the project; the availability of the rights to use the synfuels technology; water rights; plant-site rights; and the type of engineering work done to date.

During the second phase of review—the strength phase— the corporation examines additional data about the proposed technology; how federal and state environmental and safety regulations are being met; and if the sponsor has expertise in managing the project through construction and operation. In addition, the corporation wants to know if the project is economically sound and if there is a market for the final product.

## One, two, three . . .

When the corporation's new board of directors finally completed review of the applications from the first solicitation in January 1982, none of the applicants received an award. However, the board also initiated a second solicitation, which closed in June 1982. Several of the projects that participated in the first go-round submitted proposals again. Of the 35 proposals in the second solicitation, six remained in consideration after the two review phases, but none of the finalists plan to develop oil shale.

'We negotiated with the six finalists,' reports Hutchison, 'to arrive at the terms and conditions of the awards. When terms were reached, a letter of intent was issued that states that the corporation and the sponsors are pleased with the negotiations thus far, and if certain issues are resolved, we intend to provide financial assistance.' So far, one of the projects from the second solicitation, the Coolwater Coal Gasification Project, Daggett County, CA, has received a commitment from the corporation for $120 million in price guarantees.

A third and final general solicitation closed in January 1983. Forty-six projects applied, and of those, 13 were oil shale ventures. Hutchison explains that 'Unlike previous solicitations, where all of the projects were evaluated at one time, projects are being moved through the process at varying speeds. However, no awards will be made until all of the qualified projects in a resource area have been evaluated on the basis of the second stage of the review process, strength.' Because several projects need to conduct pilot plants before the strength evaluation can be completed, no final awards are expected before this fall.

Five Colorado and Utah shale projects are still under consideration in the third solicitation: Cathedral Bluffs Shale Oil Co. in Rio Blanco County, CO (federal tract C-b), which has received a letter of intent; Phase II of Union Oil Company's Parachute Creek Shale Oil Program, Garfield County, CO; Geokinetics' Seep Ridge Oil Shale Project in Uintah County, UT; White River Shale Project in Uintah County, UT (federal tracts U-a/U-b); and the Paraho-Ute Project in Uintah County, UT.

## The oil shale special

Fewer shale oil projects submitted proposals during the general solicitations than the board hoped, so plans were made to single out shale developers for awards under a separate solicitation. 'Given the fact that we have limited funds,' says Hutchison, 'the early phases of our efforts have been directed at achieving the diversity goal—that is, having many types of technologies supported by awards from early solicitations. We wanted to reach as many of the eligible resources as our funds would allow.

'We knew we'd need additional shale projects to meet our diversity goals, so the board directed the staff to prepare the Competitive Solicitation for Oil Shale Projects, which was issued this past January,' she says. The maximum award was set at $1.6 billion, and only one project was expected to receive the award. This differed from the general solicitation, where several projects negotiate for portions of the corporation's available funds.

The solicitation closed on March 15, and four of the six projects that submitted proposals were selected as "qualified bidders": Cathedral Bluffs Shale Oil Co., Phase II of Union Oil Company's shale project, White River Shale Oil Project and the Syntana-Utah Project in Uintah County, UT. Each of the qualified bidders was then required to submit a detailed technical proposal and competitive bid by June 1, 1983. Theoretically, the project that submitted an acceptable technical proposal and the lowest competitive bid would receive an award of assistance in the amount of that bid.

However, by June 1, only Union had submitted a technical proposal and competitive bid. Syntana-Utah notified the corporation that it was unable to prepare its proposal on time, and White River and Cathedral Bluffs chose to concentrate their efforts on the third general solicitation. Thus, Union was the only project left in the solicitation when the Synfuels board of directors met on June 30.

When Union's bid was examined during the board meeting, reports Hutchison, the board declared the bid "nonresponsive." This means that the bid that Union submitted did not fit the criteria that the board established for the solicitation. Therefore, no awards were made in the targeted solicitation, and the funds earmarked for distribution under the targeted solicitation remain in the corporation's coffers.

'But,' says Hutchison, 'there is still a chance that Union, White River or Cathedral Bluffs may receive an award. All three projects are still active in the third general solicitation. Also, the projects have more flexibility in the third solicitation, because the production/dollar limits set for the targeted solicitation don't apply there.'

So, 3 years after the corporation's inception, the Synfuels Corp. and oil shale are still trying to get together. But although the financial boost couldn't hurt, the shape of the industry's future still rests on a number of other factors, the world supply and price for petroleum included."

In a sidebar entitled "Some Help from DOE", the magazine explained some more of the complex dealings between the federal government and oil shale developers as follows:

"While the U.S. Synthetics Fuels Corp was getting organized in 1980, the U.S. Dept. of Energy (DOE) was authorized to make awards to synfuels project sponsors during the interim. The U.S. Congress appropriated $2 million for this purpose and put the funds under DOE's control.

DOE granted up to $400 million in price guarantees to Union Oil Company of California for Phase I of its shale project in Garfield County, CO. Union is scheduled to begin the nation's first commercial operation with a production of 10,000 barrels of shale oil per day in late 1983. Union will sell 7,000 barrels per day of diesel fuel and 3,000 barrels per day of jet fuel to the U.S. Dept. of Defense (DOD) at market price. If the market price exceeds the contract price, indexed for inflation, Union will not receive additional payments. But, if the market price is below the contract price, indexed to inflation, the Synfuels Corp., to which the Union project was transferred in February, 1982, will pay Union the difference."

The federal government's role in oil shale was well meant, but had the overall effect of slowing, if not stopping actual development. The only project, which actually produced shale oil after 1980, was the Union project. I credit them with skillful and timely negotiation with federal bureaucrats, which enabled them to produce significant quantities of hydrocarbons.

# CHAPTER 12

# The Synfuels Corporation

IN THE MEANTIME, the C-b project slowly moved ahead. Management decided that success lay with subsidization by the Synthetic Fuels Corporation (SFC). Our efforts turned entirely to developing project plans, estimates and schedules which would appear attractive to the SFC. Our contractors were more than happy to continue to re-do project plans and provide fancy proposals for the government. Discussions proceeded at a lively pace with various people from the SFC. After several iterations, an agreement was reached between C-b management and the SFC. A "Letter of Intent" was signed July 28, 1983 by Edward Noble, Chairman of the Board of the SFC, J.L. Ketelsen, Chairman of the Board of Tenneco, Inc and Armand Hammer, Chairman of the Board of Occidental Petroleum Corporation. Corporate lawyers described the document as follows:

"Occidental, Tenneco and the SFC have entered into a letter of intent dated as of July 28, 1983 concerning SFC financial assistance to the Project, to which is attached a term sheet setting forth the general terms and conditions of such assistance. The following description of the letter of intent and term sheet is qualified by reference to those documents, copies of which are attached.

The letter of intent is not a binding obligation, but an expression of mutual intent. The letter of intent states that it has been executed and the term sheet has been prepared based upon information, data, projections and estimates which were available at the time of preparation and execution.

The term sheet describes the Project and Cathedral Bluffs, SFC financial assistance and the estimated construction schedule and eligible project costs. The term sheet states that eligible project costs are estimated to be $2,665 million and provides for preparation of a further estimate of project cost. Occidental believes $2,145 million is a more likely estimate of eligible project cost, and Cathedral Bluffs submitted that lower estimate to the SFC. Given the lack of a detailed cost estimate and the difficulty of increasing the financial assistance after the agreements with the SFC would have been signed, a higher estimate was used in the SFC term sheet. Projections of financial results of Cathedral Bluffs based on the estimate of eligible project costs, which Occidental considers more likely, are used herein.

The term sheet states that new equity will be provided by Tenneco (50%) and Occidental (50%) subject to adjustment pursuant to a condition which states that the execution and delivery of the agreements providing for SFC financial assistance are subject to the following: "By September 30, 1983 unless extended by the SFC, an entity or entities satisfactory to the SFC and the other Sponsors, shall have agreed to assume a substantial share of Cathedral Bluffs and shall have executed a supplement to the Letter of Intent incorporating this Term Sheet.".

The term sheet describes the financing plan. The term sheet projects the Project's capitalization as of the date (the "Cut-off Date") which is the earlier of the date of commencement of commercial operation or March 31, 1990. This capitalization includes $373 million of new equity and $358 million of pre-Cut-Off Date operating surplus. The term sheet provides that new equity is subject to adjustment upward to the extent that pre-Cut-Off Date operating surpluses are insufficient to cover eligible project costs not otherwise covered by sunk equity, the maximum loan guarantee and any co-financing debt (as defined). Occidental believes that $207 is a more likely estimate of pre-Cut-Off Date operating surplus and Cathedral Bluffs submitted that lower estimate to SFC. The term sheet provides that upon an election to proceed after receipt of a detailed estimate of eligible project cost, the amount of the loan guarantee commitment will be reduced by 68% of the amount, if any, by which the election cost estimate is below the $2,665 million estimate used in the term sheet. The term sheet requires Cathedral Bluffs to continue to fund with 100% equity expenditures until the later of the construction election date or the time at which cumulative eligible project costs equal $315 million.

The term sheet describes the SFC price guarantee. The initial amount of the SFC price guarantee is to be increased on a dollar for dollar basis as SFC guaranteed debt is repaid. The aggregate amount of price guarantees is limited to $2,190 million. The amount of price guarantees is not to be reduced as a result of the election cost estimate being lower than the estimate used in the term sheet. The guaranteed price is $60 per barrel of hydro-treated shale oil. The SFC's obligation to make price guarantee payments will terminate ten years from the "Start of Production" (as defined).

The term sheet provides that there shall be no distributions to the partners by Cathedral Bluffs before the Cut-Off Date and that after the Cut-Off Date such distributions may be made only so long as the conditions set forth in section VI of the term sheet with respect to debt protection fund, mandatory prepayments and profit sharing are satisfied, no event of default or incipient event of default is continuing and Cathedral Bluffs has adequate working capital. At the time of each proposed distribution, an amount equal to twice the distribution is to be deposited in a debt protection fund provided that the balance in the fund will not at any time exceed the next twelve months' interest and 10% of the outstanding principal of the SFC guaranteed debt.

During the period from mechanical completion of the modified-in-situ retorts until the end of sixteen years after the Cut-Off Date, Cathedral Bluffs will undertake no "Capital Expenditures" (as defined) in excess of $20 million as adjusted for inflation, in any twelve month period without the prior consent of SFC, and during such sixteen year period Cathedral Bluffs may not increase the number of MIS retorts in primary production beyond four without SFC consent.

The term sheet includes a provision for substantial mandatory prepayment of the SFC guaranteed debt and substantial profit sharing payments to SFC out of "After Tax Cash Flow" defined to include tax benefits which would be available to a sponsor which was fully taxable.

The term sheet includes a limit on Cathedral Bluffs' indebtedness, a provision that the obligations of Cathedral Bluffs to the SFC shall be secured by a fully perfected first lien on substantially all property of the Project, and a limit on transactions by Cathedral Bluffs with affiliates.

Cathedral Bluffs may elect not to proceed with the construction of the Project after receipt of the detailed cost estimate if certain conditions exist. After election to proceed with construction and prior to the Cut-Off Date, Cathedral Bluffs may terminate its participation in the Project if it reasonably and in good faith determines at that time that, after due consideration of the economic feasibility of the Project (and not of the economic position of any Sponsor), a continuation of the Project would not be a prudent business decision. An election not to proceed with construction or a determination by Cathedral Bluffs to terminate participation in the Project may be made only upon the vote of sponsors owning in the aggregate at least such minimum interest in Cathedral Bluffs as agreed upon. The SFC has indicated that it will insist that the minimum interest be at least 66 2/3%. Upon satisfaction of the conditions to election not to proceed or to terminate participation, the sponsors will have no obligations to continue making equity contributions to Cathedral Bluffs; provided that certain obligations with respect to sponsor-guaranteed debt and the sponsors' guarantee of profit sharing payments shall continue.

The term sheet describes events of default, events of SFC suspension of assistance and remedies of the SFC.

The term sheet describes provisions for sponsor liability, in addition to the commitment to fund new equity and contingent commitment for insufficiency of pre-Cut-Off Date operating surplus (described above) and up to $100 million of cost overruns. The sponsors are to make representations concerning themselves and their "belief" (as defined) with respect to the Project. Each sponsor is to covenant with the SFC with respect to matters pertaining to its financial relationship with Cathedral Bluffs which are within its control, including the obligation to observe the limitations on distributions, permitted loans and transactions with affiliates. Each sponsor is to covenant not to take "willful actions" "attributable" (each as defined) to the respective sponsor which contribute to the right of Cathedral Bluffs to abandon the Project. Each sponsor is to covenant to notify the SFC if it forms a "belief" that either there is no reasonable assurance of timely repayment of SFC guaranteed debt, Cathedral Bluffs has the right to abandon the project or similar matters. Liability of a sponsor with respect to certain of its representations on the Project will terminate three years after the first authorization by the SFC of the issuance of SFC guaranteed debt. This time limit does apply to the covenant to give notice of lack of assurance of repayment. The liability of a sponsor for misrepresentation about itself and its subsidiary, which is a Cathedral Bluffs partner, and breach of its covenant with respect to its financial relationship with Cathedral Bluffs is not limited. The liability of a sponsor for misrepresentations about the Project and breach of the covenants on contribution to abandonment and notice of not reasonable assurance of repayment is limited to the sum of such sponsors remaining equity commitment and such sponsor's share of $100 million. Each sponsor will guarantee up to $5 million the obligation of Cathedral Bluffs to indemnify the SFC for claims by third parties arising out of the Project.

The term sheet prohibits transfer by a sponsor of its interest in the Cathedral Bluffs partner, which is its subsidiary, or the Cathedral Bluffs partnership unless the transferee meets criteria to be agreed upon and assumes all the obligations of the transferring sponsor.

No sponsor may merge into or consolidate with any other entity or dispose of more than fifty percent of its assets, unless such event will not materially adversely affect such sponsor's abilities to meet its obligations under the agreements with the SFC, the surviving entity or transferee in such event assumes all such sponsor's liabilities to the SFC and no event of default exists. The term sheet states that the agreements with the SFC will provide the procedures necessary to demonstrate that the foregoing criteria are met prior to any such merger, consolidation or disposition.

The term sheet provides for an annual guarantee fee to the SFC of 1/2 of 1% per annum of the outstanding principal amount of SFC guaranteed debt and a one-time administrative fee of $5.475 million payable one-half upon execution and delivery of the agreement with the SFC and one-half upon the first issuance of any SFC-guaranteed debt or, in the event Cathedral Bluffs elects not to proceed, upon notice of such election. Payment of the administrative fee is to be guaranteed by the sponsors.

The term sheet states conditions to execution and delivery of the agreement with the SFC including the conditions concerning finding a third sponsor described above and receipt of a favorable ruling concerning certain tax matters."

It is very important to realize that this defined project costing upwards of $2 billion was not the original 50,000 barrel per day plant defined in the Detailed Development Plan or its revisions. The project as defined in the SFC term sheet was intended to produce approximately 14,000 barrels per day of a commercial crude oil substitute. The project included mining, a commercial aboveground retort using Union Oil Company's Unishale B process, four continuously burning modified in-situ retorts using the MIS technology developed by Occidental Oil Shale, Inc., and upgrading facility for treatment of raw shale oil, an approximately 46 mile pipeline and terminal and incidental facilities. Oil shale feedstock to the aboveground retort would be provided primarily by ore from a commercial room and pillar mine and supplemented by ore obtained from preparation of the MIS retorts. The crude shale oil would be upgraded on the tract (using Union Oil Company's upgrading technology) to produce a substitute crude oil. The upgraded synthetic oil would be shipped via pipeline to a distribution terminal in Rangely, Colorado. The project was expected to produce approximately 11,000 barrels per day of raw shale oil from the aboveground retort and approximately 2,300 barrels per day from the four MIS retorts. The oil upgrading facility would produce approximately 14,100 barrels per day of a commercial substitute crude oil. (The additional 800 barrels a day would come from hydrogen added to upgrade the crude oil. The hydrogen basically comes from breaking down water into its component parts.) The expected production life of the project was not less than thirty years producing approximately 150 million barrels of upgraded synthetic crude.

The careful reader has noticed the introduction of a new theme in the SFC letter of intent. There are provisions for new private partners, which may be courted by Oxy and/or Tenneco. With this agreement in hand, we did indeed go out into the investment world and try to find others who might be interested in sharing in the risks and potential rewards of the C-b Project. Large corporations were approached and they listened to the proposal, but no one agreed to the proposition. These sad realizations, e.g., a 14,000 barrel per day facility instead of 50,000 barrels, capital and operating cost predictions expanding beyond belief, capped by the awareness that thoughtful profit-oriented corporations shunned the idea of joining the partnership, signified the beginning of the end of the C-b project.

# CHAPTER 13

# Beginning of the End

IN A SPECIAL report, Shale Country published an excellent accounting of one of the major events in the shale oil development process as follows:[xi]

"On Sunday, May 2, 1982, Exxon Company, U.S.A. announced that after reexamining costs, the Corporation's board of directors had decided to cease financing Exxon's portion of the Colony Shale Oil Project located near Parachute, CO. Because of that decision, the other Colony owner, The Oil Shale Corp. (a Tosco subsidiary), exercised an option it held in the agreement and sold its 40-percent share of the project to Exxon. SHALE COUNTRY spoke with Bob Larkins, manager of the Synthetic Fuels Dept. for Exxon, to find out how Exxon arrived at its decision and what happens next.

S(hale)C(ountry): Why did Exxon decide to stop funding Colony and how did the decision-making process work?

**Larkins:** Periodically we review all of our large projects and we examine all aspects of a project's attractiveness. We look at such things as investment and operating costs, and we put in a dollar allowance for adjustments and changes. We also look at how rapidly a project can be brought to full capacity. With Colony, the change in investment costs was the principal factor in our decision—the economics were no longer attractive enough. Our forecasts showed costs of the plant rising substantially—$5 billion or $6 billion versus the $2 billion or $3 billion estimated 2 years ago.

This investment change was the main reason for ceasing to fund Colony, because other factors in the long-term external environment had not changed. For instance, the price of oil is down, but this is a near-term change and did not play much of a part in our decision. The decision was based on expectations at start-up, 5 or 6 years from now, and for operating life of the plant after that. The recession was not really a consideration either, because from the project's point of view the cost escalation would have been less.

**SC:** The original cost estimate for Colony was approximately $3.2 billion and now projections are running as high as $6 billion. Why is there such a large difference between the two estimates?

**Larkins:** When you estimate the cost of a refinery, you know exactly what is required—they've been built many times before. With oil shale, whether it's Tosco or Union technology, the plant is being built for the first time. Because Colony was a first-generation plant, we were finding more and more things that needed to be done, or done differently, or that hadn't been thought about earlier. And all of these factors were adding to the cost of making the plant reliable, operable and safe.

**SC:** Exxon, of course, must have been studying Colony's costs continuously, but when did costs start to become a serious concern?

**Larkins:** There's really no one time that I can point to. Perhaps the real concerns started with the detailed project review that we did in November 1981. Then, in the beginning of 1982 when we began reviewing the project again, we could see no relief—even after examining alternative development plans. The concern was that more work was not likely to lead to a better answer. And money was being spent at a fairly high rate.

We wanted to alert Tosco in a timely fashion, and we had alerted both Tosco and the Synthetic Fuels Corp. to some degree back in March. We indicated that Exxon was in the midst of serious project reviews that could result in a range of decisions from continuing Colony to ceasing to fund it. Then, when the board of directors met on Wednesday, April 28, it reviewed the project's status and outlook, and the decision was made. We notified Tosco of the board's decision to discontinue funding the project on Friday.

**SC:** Why was the board's decision announced so abruptly? Was there another way the announcement could have been handled?

**Larkins:** Exxon management communicated the board's decision to Tosco on Friday, April 30th. Tosco had a number of alternatives it might have considered and we were anticipating a discussion of these alternatives. For instance, it had a right to reduce the size of the project from six retorts to two.

It had a month in which to consider alternatives, but instead, after a fairly short period of reflection—over the weekend, in fact— Tosco's board reviewed the situation and decided to exercise the option to sell its 40-percent share to us. Once that decision was made the situation took on more sensitivity and the announcement came quickly. Because of Tosco's stock-market sensitivities, we had no choice but to make the announcement before the stock market reopened on Monday. We didn't have a chance to give advance warning to state and local officials, or even our own people.

**SC:** Exxon has said it will close down Colony in an orderly manner. What does this mean?

**Larkins:** There has been a lot of press coverage implying that every last Exxon person is going to disappear from the Western Slope, but this just isn't true. We think the oil shale resource is important to this country's future. The project is in mothballs rather than abandoned.

We will continue to have a small number of people working at the site. Some things will have to be maintained, such as the roads and the mine bench. And, even if we were not expecting to resume the project at a future date we would be obligated to reclaim and revegetate.

There will also be people who will remain to handle Battlement Mesa (the new community that was being built by Exxon and Tosco subsidiaries). As you know, we have decided to continue to develop this community on a reduced scale. Utilities, roads and essential services for the dwellings under construction will be completed. The rate of completion of housing units, public facilities and commercial developments will proceed in accordance with demand. We will be working closely with affected community and other officials involved, both with regard to the pursuit of an environmentally acceptable reclamation plan and the continuation of Battlement Mesa.

**SC:** Exxon has said that it might reactivate the Colony Project at a future date. What would be necessary for this to occur?

**Larkins:** It's hard for me to predict what types of things might be necessary. Things have not been smooth with oil prices in the past and we might expect that there would be bumps and jolts in the future for both supply and prices. We think the Tosco technology is the best currently available for a full-sized commercial plant, but there are other technological alternatives under development. If one or more of these were successful, we would certainly consider them, and thus the project might not be reactivated in its current form.

**SC:** Some industry observers have suggested that Exxon's decision to withdraw from Colony means the end of synthetic fuels development in general and oil shale in particular. Do you agree with this theory?

**Larkins:** No, we still feel that the United States needs synfuels to reduce its reliance on foreign imports. When Exxon published its white paper in 1980 (possibility of an 8-billion-barrel-per-day shale oil industry early in the 21st century), the paper outlined three questions. 1) What is possible with respect to this resource? We don't feel the 8-billion-barrel-per-day possibility has changed. 2) How much is needed? The forecast of volumes needed and timing have changed. 3) What is likely? This is the most difficult question and addresses all the very real problems—such as cost and economics—in order to respond to the question of what is likely. The pace of development will be slower than we thought a couple of years ago, but we feel the need for synfuels will continue to be there in the long-term.

We made the decision to stop financing the Colony Project reluctantly. We appreciate the special significance this project had and took that into consideration during our reviews. Exxon continues to believe in the long-term need for synthetic fuels and wants to be in on their development. It was not an easy decision and, as I said before, it was made reluctantly."

In dramatic contrast to the above 1982 explanation by Exxon of their Colony shutdown decision is the "white paper" referred to by Mr. Larkin. In 1980, Exxon prepared a paper for national study and reference explaining their corporate position on synthetic fuels. This paper is reproduced herein as follows:

*Synthetic Fuels: A National Critical Need*

Exxon's continuing analysis of energy trends conducted in the regular course of business planning has led to a growing conviction that rapid development of a synthetic fuels industry in the United States is a critical national need.

A synthetic fuels industry can start contributing to domestic energy supply by the end of the 1980s and can grow to a large sustainable volume in the '90s. Several processes for making synthetic fuels are in advanced stages of development and are ready for application in commercial-sized plants.

Such processes can apply the country's greatest energy strength to its most pressing energy need, by converting abundant domestic resources—oil shale and coal—into liquid and gaseous forms for uses now served by oil and natural gas.

Timely development of a synthetic fuels industry is essential to any realistic expectation that the country's need for imported petroleum can diminish substantially in this century.

This development is critically important to national security and to standards of living in the future.

The synthetic fuels industry can represent a substantial strengthening of the country's industrial base. Its growth potential offers some of the most significant business and employment opportunities in modern history.

But development on a timetable consistent with national security and economic needs will require cooperation of a degree that may be unprecedented in peacetime among business, labor, government at all levels, and other interest groups.

The public understanding requisite for such cooperation does not exist today. This report, intended to contribute to the needed awareness, presents Exxon's assessment of the opportunities available to the United States for meeting future energy needs with emphasis on the role of synthetic fuels. Data projections and illustrations are drawn from a U.S. energy outlook and other studies that are developed by Exxon and regularly revised in the light of events, as a basis for internal planning, corporate strategy and investment decisions.

## An Energy Transition Already Begun

A domestic synthetic fuels industry has a potentially important role but cannot provide a total solution to America's energy problems. It will not eliminate needs for aggressive efforts in development of other forms of energy and in energy conservation. It should be viewed in a realistic context as one of several tools essential to successful management of an inevitable energy transition that already has begun and will continue for generations.

The U.S. and world economies are shifting from dependence

on energy provided principally by oil and gas to a more diversified mix of supplies. The transition will likely lead, during the 21st Century, to substantial reliance on energy from renewable and nondepleting sources.

For the United States, the need for a transition to other energy sources began to develop when the country started producing oil and natural gas faster than new reserves could be discovered. Domestic petroleum production is declining. and Americans have become dependent on foreign countries for almost half of the oil they use. The rest of the free world is now following the pattern—using more petroleum than is being found. Production in some of the Middle Eastern countries with the world's largest oil reserves eventually will plateau and decline—perhaps in the early part of the 21st Century.

The transition to non-petroleum forms of energy will continue for decades, but some of the difficulties involved are already apparent.

U.S. consumers have experienced temporary shortages of oil products on several occasions—including those resulting from the embargo of 1973-74 and the revolution in Iran in late 1978. The balance of world oil supply and demand remains precarious. And for years to come, the risk of shortfalls is likely to be greater than the prospect of surpluses.

The rising price of oil imports, and the country's increasing dependence on them, have contributed to trade-balance problems, inflation, declining real personal discretionary income for individuals, and other economic difficulties. An end to the trend of rising real costs of energy is not in sight.

A crucial question is whether the nation will continue to "muddle along," or will begin to successfully *manage* its way through the transition—seizing its feasible economic opportunities to save energy and to develop additional domestic energy supply. Unless this happens, succeeding generations of Americans face the prospect of real decline in economic well being, reduced opportunities for employment and advancement, and further loss of independence in international affairs.

There are numerous opportunities to improve the domestic energy balance. But they are not, generally speaking, alternatives to one another. Some can affect energy demand or supply almost immediately. Others can produce benefits within the next decade or two. Still others will have no significant impact until well into the 21st Century. Furthermore, not all forms of energy can substitute for the scarce fuels—conventional crude oil and natural gas in all uses.

Failure to make distinctions concerning the time frames and the uses in which potential sources of energy can help could lead easily to planning based on wishful thinking—with potentially disastrous consequences for U.S. security and long-term economic health.

## Energy Conservation

The most obvious widely recognized opportunity to improve the domestic energy balance—and one that can have immediate impact on the level of petroleum imports—is energy conservation.

Because lead-times required to develop new energy supplies generally range between six and 12 years, near-term domestic supply is essentially fixed—leaving reduction of demand as the only short-term way to lower significantly the U.S. need for imported energy.

Consumers throughout the economy are responding to the rising costs of energy by finding ways to use it more efficiently—in homes and stores, in transportation, and in industry. In the operations of Exxon USA, for example, an intensive conservation effort is saving enough energy to meet the electricity needs of more than one million American homes.

Savings also are resulting from better new car gasoline mileage, lower highway speeds, thermostat adjustments and other actions mandated by government.

The combined effect of voluntary and mandatory improvements in efficiency is to hold energy consumption well below levels that would have been reached if price and use patterns had followed the trends experienced prior to the 1973-74 embargo.

U.S. energy demand this year is about 13 percent below the "trends-continued" level—reflecting a savings of six million B/DOE. Savings are projected to reach 27 percent, or 16 million B/DOE by 1990 and 35 percent, or 27 million B/DOE by 2000.

Improved efficiency also has altered the relationship between energy use and economic growth. During the 13-year period preceding the Arab oil embargo. it took a 4.1 percent average annual increase in energy consumption to support an identical 4.1 percent rate of growth in real Gross National Product. But it is projected that in the 1980s and '90s economic growth will average 2½ to 3 percent while energy consumption grows at rates averaging only 1 to 1½ percent.

It would not be prudent however, to rely solely on conservation as a solution to energy problems. Despite improving efficiency, some increase in energy consumption still must occur if the economy is to grow sufficiently to meet employment and income needs of a growing population.

At the projected 1 to 1½ percent average annual growth rate, U.S. energy demand in the year 2000 would be about 30 percent higher than it is today.

## Sources of Energy Supply

America has substantial energy resources, which potentially can help meet this demand. Eventually their development can enable the country to start reducing its need for foreign petroleum—although oil and gas imports are expected to continue to be significant at least through the mid or late 1980s.

Each feasible source of additional supply has its capabilities and limitations. Each can supply certain quantities of energy in certain time frames, in forms suitable for certain uses.

Non-depleting and renewable sources are not expected to meet substantial portions of energy demand until well into the 21st Century.

*Hydroelectric* and *geothermal* energy, for example, are limited by the availability of suitable sites.

*Solar* energy offers significant potential, but its growth is presently inhibited by high costs. Converting the sun's rays directly into heat with today's technology is typically three to four times as expensive as obtaining the heat from conventionally generated electricity. Technological breakthroughs eventually may bring costs down, and should be pursued vigorously. But it should be recognized that, even after the breakthroughs occur, it will take decades to apply the new technology widely enough to meet a significant share of energy demand.

Solar units for space and water heating are becoming more competitive in areas where electricity costs are high. The rate of deployment, however, is likely to be slow. Even if economics could be ignored and such units could be installed in *all* new homes, beginning today, more than 20 years would pass before the units could meet 1 percent of the country's energy needs. (This assumes that the units would provide one-half of the new homes' space and water heat. The other half would come from conventional back-up heating systems which take over when sunlight is inadequate.)

Direct conversion of solar energy into electricity is even more expensive and farther away from economic feasibility.

*Nuclear fusion* ultimately may add dramatically to the already substantial potential of nuclear technology to provide electricity. But fusion is still in the research and development stage and commercial demonstration is unlikely before the year 2000. Even after commerciality is demonstrated, fusion will not displace other sources overnight; it will be phased in as electricity demand grows and older plants become obsolete.

Small quantities of liquid fuel are now being provided by ethanol, an *alcohol* made from grain or other crops and then blended with gasoline to form gasohol. Costs of the ethanol are higher than those of foreign oil or of synthetic fuels that can be derived from oil shale or coal, but this fact is masked by tax exemptions and other subsidies. Alcohol does have some potential for contributing to domestic liquid fuels supply—provided producers turn from oil and natural gas to coal or biomass as sources of the energy they use in processing. U.S. and world food requirements place limitations, however, on acreage that can be used to grow alcohol feedstocks. And these limitations suggest that alcohols made from renewable crops can have only modest long-term impact on the energy transition in this country.

During the decade of the 1980s, virtually all growth in domestic energy supply will have to come from conventional nuclear power and from direct burning of coal.

*Nuclear fission* has been the nation's fastest growing source of energy. It now meets about 5 percent of energy demand, a share that could increase to 13 percent by the year 2000 if concerns about safety and other problems are satisfactorily resolved. Enough reactors are under construction or on order to meet this projection through the 1980s. But the ordering rate will have to increase soon, reversing the trend of the past few years, if projected growth through the 1990s is to be achieved.

Growth in the use of *coal* over the next few decades will be limited only by demand. Its conventional uses now account for about 19 percent of energy consumption, and this share is expected to increase to about 33 percent by 2000.

To achieve this growth—and to provide coal for feedstock in making synthetic fuels—mining production will have to increase from about 730 million short tons this year to almost 1.3 billion in 1990, and to more than 2.2 billion in the year 2000. Because coal reserves and the mining industry's expansion capabilities are more than adequate, this growth is a realistic, achievable objective.

It should be emphasized, however, that nuclear power and coal have only limited usefulness in meeting some significant needs now served by *oil* and *natural gas*.

Nuclear provides energy only as electricity. And coal, with today's technology, substitutes for oil and gas only as boiler fuel for electric generators and other large industrial facilities. Conversions of boilers from petroleum fuels to coal often are very costly, and much of the economically feasible conversion already has occurred.

There are large segments of energy demand—particularly for transportation and petrochemical feedstocks—that can be met only by liquid and gaseous fuels. Current U.S. demand for such fuels, about 28 million B/DOE is projected to decline only gradually—remaining above 25 million B/DOE through the rest of the 20th Century. These segments of demand will have to be served by domestic petroleum imports and synthetic fuels.

Future discoveries of domestic oil and natural gas will be critically important. They are expected to account for about 30 percent of the domestic petroleum production of 1990 and over half the production projected for 2000. This dependence on new discoveries emphasizes the need for a continuing high level of exploration activity, and for leasing unexplored federal lands and offshore areas that may have geological promise. Every new barrel of oil or gas production the nation achieves will directly reduce its need for imported energy.

U.S. oil and gas production has been declining, however, since the early 1970s. Virtually all new reserves will be difficult and expensive to find and develop. Frontiers of exploration are moving into Arctic environments and deeper ocean waters. And, under the best of circumstances, major discoveries made in the next few years can come into full production no sooner than the late 1980s.

In light of these factors and the disappointing results of exploration over the past decade, it seems likely that new discoveries will continue to be less than production. The only realistic planning assumption is that domestic petroleum production will continue to decline.

Even with substantial conservation efforts, this points to a continuing high level of *petroleum imports* through the 1980s, and to a continuing need for working relationships with oil-exporting nations. Governments in the producing countries understandably want to conserve their resources and are reluctant to meet new increases in demand.

## Key Role of Synthetic Fuels

It may be possible, however, for the United States to significantly reduce its dependence on imports in the 1990's-provided planning and construction of *synthetic fuels* plants begin now and accelerate rapidly within the next few years. It is projected that these fuels' share of total U.S. energy demand can reach 12 percent by 2000 and will continue to increase in the 21st Century.

Demand for oil and gas is expected to decline gradually as substitution of other fuels, coupled with technological breakthroughs such as improvements in the electric automobile, allows non-petroleum sources of energy to meet an increasing share of total domestic energy needs. A substantial demand for liquid and gaseous hydrocarbons is expected to continue, however, well into the 21st Century.

Potential demand for synthetic fuels—the difference between projections of domestic petroleum demand and production—is likely to reach an estimated 15 million B/DOE early in the 21st Century.

Obviously these projections are subject to variation, but the essential point is that for decades to come there will be a large demand for energy in liquid and gaseous forms which cannot he satisfied with domestic oil and natural gas. In meeting this demand synthetic fuels are the alternative to continued reliance on imported petroleum through the 1990s and beyond.

If new domestic discoveries of oil and gas should turn out to be greater than expected, and if the nation also has moved rapidly to develop its synthetic fuels potential, the variance from projections would be beneficial—allowing faster progress in reducing the nation's dependence on imported petroleum.

On the other hand, exploration results may fall short of projections. If they do, and if progress in synthetic fuels development also is disappointing, the resulting increase in need for imports would be a problem of serious proportions. In view of the present precarious balance of world oil supply and demand

and the future outlook, it is unlikely that the needed quantities of foreign petroleum would be available.

Among the potential supplements to domestic oil and gas supplies, synthetic liquids and gases made from oil shale and coal have potential, in contrast to the alcohol fuels made from grain, for build-up to large sustainable volumes. A number of processes for making fuels from coal and oil shale have seen successful commercial use in other countries. Others have been in research and development stages for years, and are approaching readiness for commercial application.

*Shale oil* now appears to be economically competitive with imported crude oil. It is obtained from oil shale—a finely textured sedimentary rock containing an organic substance called kerogen. The rock can be mined. with underground or surface techniques, then crushed and heated to yield an oil resembling heavy crude. After processing to remove impurities and lighten the oil, conventional refining methods can turn it into a variety of oil products—including gasoline. jet fuel, heating oil and fuel oil.

Several methods of processing oil shale in large *surface retorts* have been demonstrated in large pilot plants and appear ready for scaling up to commercial size. Efforts to find feasible ways of using *in-situ* techniques, which would process the shale underground. are continuing—although early attempts were generally unsuccessful.

An *intermediate BTU gas* (IBG) made from coal also appears to be economically competitive. IBG is a mixture of hydrogen. carbon monoxide, and methane. When burned it yields about 40 percent of the heat that could be produced by burning a like volume of natural gas. It is suitable for use as a fuel for large industrial facilities and as a valuable feedstock for petrochemical processes, but not for mixing into conventional natural gas transportation systems.

The heat content of IGB can be increased, by a process that converts the hydrogen and carbon monoxide into methane, to provide a *synthetic natural gas* (SNG). It is indistinguishable from—

and could be mixed, transported and burned with natural gas. This additional processing increases costs by some 15-25 percent, but SNG potentially can supplement gas supply for homeowners and specialized industrial uses now requiring natural gas.

The least costly method of obtaining *liquid fuels from coal* with today's technology appears to be an indirect process for producing *methanol.* The coal is first processed into IBG, which in turn is further processed to yield both methanol and SNG. The methanol can serve many traditional fuel and chemical uses. Its cost—with appropriate adjustments for differences in location and product qualities—is estimated at about 20-30 percent higher than shale oil, IBG, or imported crude oil. But this estimate is sensitive to the revenue the producer realizes from the sale of the co-produced SNG.

Much research and development work is being done on other technologies, which convert coal, directly or indirectly, into other liquid fuels. Costs of these liquids are now estimated at 40-60 percent above IBG, shale oil or imported crude oil, but technological advancements are likely to bring costs down. These processes will be needed to allow efficient utilization of a wider range of coals, but are not yet ready for commercial application.

Much synthetic fuels technology, however, is now ready for commercial use. And the domestic resource base is adequate to support a very large industry.

*Resources and Investment Needs*

Known recoverable reserves of coal and oil shale even after deducting coal to be used conventionally and the energy to be consumed in processing—are capable of providing synthetic fuels equivalent to one trillion barrels of oil. That's three times as much energy as the U.S. Geological Survey estimates can be provided by the country's remaining proved and undiscovered reserves of oil and gas. And it's enough to sustain a synthetic fuels industry producing 15 million B/DOE for 175 years.

The investment required to attain production of 15 million B/DOE is staggering—almost $800 billion in 1980 dollars (over 3 trillion 'as spent" dollars). This would be spread over 30 years or more, however, and is within the capabilities of private companies.

In peak years, it would represent about 1 percent of GNP— roughly what is now expended for conventional domestic exploration and production of oil and gas. To put it another way, these capital expenditures would constitute about one-tenth of private, nonresidential investment.

## A Massive But Manageable Task

Considering the readiness of technology, size of the resource base, and financial capabilities of the private sector, a goal of meeting the potential synthetics demand of 15 million B/DOE by the end of the first decade of the 21st Century would be highly ambitious— but not beyond achievement by a determined America.

If the synthetic fuels industry develops at such a speed, it will be possible to begin reducing national dependence on petroleum imports within a decade, reduce them significantly during the 1990s, eliminate them early in the next century (if their elimination still appears desirable) and assure adequate supplies of liquid fuel and gas despite an anticipated decline in world oil production.

The task will be massive, however, by any standard of comparison. And it will be complicated by the fact that much of the industry will have to be concentrated in arid, sparsely populated parts of the West.

The output of a cost-effective 15 million B/DOE industry would include some eight million B/D of shale oil from high-yielding deposits in the Piceance and Uinta Basins in Colorado and Utah. About one-fourth of the shale could be mined underground, but achieving the projected volume would require that the rest be obtained by large-scale surface mining. (Surface mining can recover up to 10 times as much shale as underground room-and-pillar methods.)

The remaining seven million B/DOE of synthetic fuels would be derived from coal—three million from the Powder River Basin of Wyoming and Montana, and another four million from other parts of Montana, the Dakotas, the Southern Rockies, the Interior Basin (Illinois and parts of adjacent states), Gulf Coast areas, and Appalachia.

## Water Needs

Shale oil and coal synthetics require from two to four barrels of water for each barrel of product. And water supply is a major concern in dry Western regions where much of the development would occur.

Unless additional water is transported into these regions, it is unlikely that the industry's production can exceed seven million B/DOE—less than half the potential demand.

It appears financially and physically feasible to bring the additional needed water to the Piceance and Powder River Basins from the Missouri and several other Western rivers—at a resulting increase in costs of about 1 percent. As a side benefit, the transportation systems also could provide water for agricultural, municipal and other uses.

## Employment and Population Effects

A synthetic fuels industry capable of producing 15 million B/DOE in the year 2010 would employ an estimated 870,000 people—480,000 in mining and 390,000 in the synthetic fuels processing plants. In peak development years, its growth would also provide 250,000 jobs in construction and 8,400 in design engineering.

Over the span of 30 years or more, the new industry would directly increase the nation's total number of jobs, over current levels by 60 percent in mining, 55 percent in process industries, 15 percent in construction, and 35 percent in the applicable design engineering specialties.

These direct employment effects would be multiplied, of course, by the industry's demand for machinery and other products from suppliers, and by growth of new population centers in the development areas.

Careful planning will be required in these areas to develop, almost literally from scratch, transportation systems, housing, school systems, health care facilities, numerous public services and other essentials needed by thousands of workers and their families.

This will be a formidable challenge, but the task is manageable because the buildup will begin slowly and will he sustained over a long period. Nonetheless, the importance of timely advance planning cannot be overemphasized.

*Summary and Conclusions*

The need for rapid development of a synthetic fuels industry—to ease dependence on imported energy and to meet a coming shortfall of world oil supply—is clearly critical to the national interest. The resources are adequate. The technology is ready for commercial application. And the private sector has the necessary financial, technical and managerial resources.

There is no need for the federal government to assume the risk of a producer. But governments—at all levels—do have major roles to play.

The needed transfer of water, for example, cannot occur without governmental involvement and broad-based public support.

Coordination by industry and by governments at local and state levels will be essential in adequately planning for population influx and other impacts in the development areas.

Perhaps the greatest responsibility of the federal government, reinforced by state and local governments, is leadership—in building the public awareness of the critical national need to develop domestic synthetic fuels production as rapidly as possible and in creating the economic and regulatory climate to facilitate the new industry's growth.

A second major responsibility of the federal government is to function more effectively in a managerial role—reconciling conflicting priorities and regional interests, and assuring that studies, hearings and permitting processes proceed without inordinate delays of the sort that have slowed energy projects so often in the past.

Without bold, responsible, bipartisan political leadership, it is unlikely that the promise of synthetic fuels can ever be fully realized. And whether or not the industry's full potential is achieved will depend largely on how well industry, labor, civic leadership, environmentalists, and the governments of cities, counties, states and the nation can work together.

Vigorous effort must begin immediately if the national interest is to be served. Plans for the country's first synthetic fuels plants are now being made by several firms. But meeting critical national needs will require that work begin now on plans for continued development, beyond these initial plants, and even beyond the 1990 goals suggested by the President in the summer of 1979. What is needed is not just a few new plants, but an entire new industry.

The task is great. So is the need. And there is no time to lose.

# CHAPTER 14

## Socioeconomic Aspects of the Oil Shale Boom

B EFORE WE GO on to describe the exciting demise of the oil shale industry as it developed in the 1980's, further discussion concerning industry and government's planned and completed projects to build the infrastructure considered necessary for the projected commercialization of a major industry in western Colorado and Eastern Utah need to be put into the record.

Cathedral Bluffs was always a leader in the efforts to make the impact of adding a very large work force to a relatively unpopulated area as smooth as possible. Our overall plans included, from the very beginning, various provisions in the project scheme, design and estimates to mitigate adverse socioeconomic impacts resulting from our development. We participated in and sponsored a number of programs including cooperative efforts with state and local governments to assess and understand all the impacts of cumulative growth and the capability of public facilities to meet demands for increased services.

Early on, C-b, along with the other federal tract lessees, had paid millions of dollars of bonus payments to the federal government for the right to develop their tracts. The provisions of the lease program called for 37.5% of the cash payments to be handed over to the state for public impact mitigation purposes. Thus, Colorado received some $47.5 million from tract C-a and $27.5 million for C-b. This money, which came to be called the Oil Shale Trust Fund, was utilized for impact mitigation under a less-than-simple system. Most population growth in Colorado was projected to occur in Rio Blanco, Garfield, Mesa and Moffat counties. County representatives early on estimated the price tag for required schools, sewers, roads, recreational areas, additional police and fire protection, and medical services at more than $1 billion for a 400,000-barrel-per-day industry. And most of this expenditure would be needed to be made *before* Colorado's oil shale counties begin to accumulate much of the tax revenue that shale industry growth could be expected to generate.

The process to get this money from the state was interesting. Each year from 1975 through 1980, the impacted counties and municipalities determined project priorities, then the Region 11 Colorado West Area Council of Governments (COG) compiled the requests and worked out the logistics. The impact financing requests would be presented to the state legislature, which approved allocations on a project-by-project basis based upon recommendations of the state's Joint Budget Committee (JBC). Planned projects for the trust fund monies included fire, school and city government facilities; senior citizen housing; hospital equipment; water systems; and street and airport improvements. Each county drew up its project list in a similar manner. In Garfield County, for example, the local citizens argued, set priorities, cut and slashed before presenting requests to the legislature. Garfield County arrived at its final list of projects through the efforts of two local committees—the West Garfield Impact Committee and the Garfield County Core Group. The first group consisted of a cross-section of professional people, senior citizens, hospital and health services groups, public safety and public works groups (such as police and fire officials), and elected officials. The recommendations of this group were studied by the Core Group, which was comprised of the mayors of the six county municipalities and the three county commissioners. The Core Group recommendations were then forwarded to the state legislature, but only after all this intensive local review and discussion.

The JBC's early plan was to invest the trust fund principal and use only accrued interest for state expenditures. Early allocations were for construction and improvements of roads, schools, water and sewer projects, and bridges. Examples of initial trust-funded projects were: street improvements for the town of Silt, costing approximately $1.3 million; improvements for the Meeker high school, $1.1 million; and an extension of Grand Junction's airport facilities of approximately $4 million. These projects were only the beginning of steps taken to prepare the Western Slope of Colorado for the population influx projected to accompany oil shale development.

Deciding who gets what at the county level was a significant learning process for local government. The system developed into a smooth and exemplary process.

In 1980, in what I've always considered to be an enlightened and unselfish move, the legislature voted to turn over more than $47 million remaining in the trust fund to the counties. The county commissioners developed their own formula for dividing up the money, by studying impact forecasts and recommendations from professional planners. They requested that the legislature apportion $22.3 million to Garfield County, $16.7 million to Rio Blanco County, $6.5 million to Mesa County and $2.2 million to Moffat County. The counties wanted the flexibility of spending allocated funds on the most appropriate projects as the oil shale industry developed and as impact needs became apparent.

Expenditures from 1975 to 1981 for Moffat County included about $2.5 million for schools, $700,000 for health care facilities, $3 million to the town of Craig, $200,000 to the town of Dinosaur and $400,000 for other county projects. Along with the $2.2 million turned over in 1980, Moffat County received nearly $9 million from the Oil Shale Trust Fund.

Similarly, during 1975 to 1981, Rio Blanco County spent about $2.5 million for schools, $300,000 for health care facilities, $3 million for the town of Rangely, nearly $5 million for the town of Meeker plus $8 million for other county projects. With the $16.7 million turned over to them in 1980, Rio Blanco County residents received $35.6 million from the Oil Shale Trust Fund.

From 1975 to 1981, Mesa County received and spent $1.1 million for schools, some $35,000 for health care, $100,000 for Grand Junction, $300,000 for Fruita. $150,000 for Gateway, $1.4 million for Debeque, $80,000 for Collbran, $45,000 for Palisade and $6.7 million for other county expenditures. Along with the $6.5 million distributed to them in 1980 by the legislature, Mesa County received some $16.3 million from the trust fund.

Garfield County received the most money of all Colorado counties from the Oil Shale Trust Fund. During the 1975 to 1980 period, the county allocated $7.3 million to schools, about $300,000 to health care, $500,000 to the town of Newcastle, some $45,000 to Glenwood Springs, $6.4 million to Rifle, $1 million to Carbondale, $3 million to Silt, $900,000 to Parachute and about $3 million for other county expenditures. When the $22.3 million was added in 1980, the total that Garfield County received from the trust fund totaled $44.8 million.

After the trust fund money was allocated, a major concern for all four counties became finding ways of investing trust fund money, to insure a reserve for the future, while meeting the most immediate county needs.

This extraordinary attention to communities and their infrastructure was not pie-in-the-sky pork-barrel spending. There were real impacts planned and indeed, occurring in western Colorado. As an indication of the kinds of things that were happening, the following is a summary of a housing study we at C-b undertook based upon our proposal submitted to the SFC. Remember, this was based upon a small, 14,000 barrel per day plant. This was a housing assessment and analysis conducted in preparation of a Cathedral Bluffs Housing Master Plan, developed to provide adequate housing accommodations for our project-related workforce.

*Existing Housing*

The assessment of the existing housing conditions revealed that the study area (Rio Blanco and Garfield Counties) had considerable capacity for additional growth. The private housing industry appeared willing to respond to projected housing requirements, however, high vacancies and an uncertain economic climate would likely reduce the availability of risk capital and construction financing for speculative development. The existing housing conditions were summarized as follows;

- The study area housing stock between 1970 and 1980 increased 4,304 units, or 56.9 percent, and an estimated 2,735 units, or 23.0 percent, from 1980 through September 30, 1983.

- The study area's nine communities contained over 2,000 vacant housing units/spaces in late 1983; of which 1,765, or 87.8 percent, were located in Meeker, Rifle, Parachute, and Battlement Mesa.

- Adequate developed, (with infrastructure), and undeveloped platted residential lots capable of supporting a large in migrating workforce existed within the study area, primarily in the Meeker, Rifle, and Battlement Mesa areas.

- The study area housing industry had the capacity to develop in excess of 500 new residential units/spaces per year.

- Current Meeker, Rifle, Parachute, and Battlement Mesa rental housing rates had declined 30 to 40 percent since 1981.

- The study area's overall capacity for residential growth was considered sufficient to accommodate in-migrating workers associated with the Cathedral Bluffs Project. The major development constraint centered around the Project's demand uncertainty and the characteristics of this demand; i.e., type of housing required, locational factors, and timing requirements.

*Housing Demand*

Cathedral Bluffs employment data were the basis for projecting annual average housing demand numbers. These projections were prepared by Cathedral Bluffs using a Cumulative Impact Task Force (CITF) PAS Model,(This model was developed with State participation to provide a generally agreed upon basis for forecasting housing and other population demand requirements.) This system, along with everything else associated with oil shale development, was controversial and engendered lots of arguments among the assembled real and so-called experts. The model was designed to distribute the housing impact throughout the study area, primarily to a mancamp, Meeker, Rifle, and the Parachute/Battlement Mesa area. The assessment of housing needs was designed to identify the specific housing demand for the Cathedral Bluffs Project temporary construction and permanent operations and induced workforces. The project-related housing demand had two distinct characteristics. The first was strictly temporary, required for the five-year construction schedule. The second was the permanent demand for housing created by the in-migrating operations and induced employees. The housing demand characteristics were summarized as follows:

- The direct construction and operations workforces reach an annual average peak of 1,874 employees in 1986 and 1,012 employees in 1989, respectively.
- The baseline housing demand was projected to decline by 58 housing units between 1984 and 1989.
- The direct construction housing demand reaches an annual average peak of 1,543 units in 1986.
- The direct operations housing demand reaches an annual average peak of 779 units in 1989.
- The induced housing demand reaches an annual average peak in 1989 of 825 units, then declines to a steady-state level of 762 units in 1989.
- The Project-related new housing demand builds to an annual

average peak of 2,638 units in 1986 before declining to a steady-state level of 1,541 units in 1989.

- The direct construction demand is projected to occur in the Mancamp (37 percent), Parachute/Battlement Mesa (33 percent), Rifle (13 percent), and Meeker (10 percent).
- Approximately 90 percent of the direct operations and induced steady-state housing demand is projected to take place in Rifle and Meeker.

*Housing Mitigation Measures*

The Housing Mitigation Measures were designed to ensure adequate temporary and permanent dwelling units to meet Cathedral Bluffs workforce requirements. The mitigation measures were presumed to be guides to aid both Cathedral Bluffs Management and the local housing industry in developing the required housing supply. Based upon existing conditions within the study area and C-b's housing demand, the following mitigation measures were identified:

- Utilizing up to 1,150 existing vacant apartments, townhomes, condominiums, mobile homes, mobile home pads and motel units, primarily in Meeker, Rifle, Parachute, and Battlement Mesa. The location and availability of units within a reasonable driving proximity of the C-b Plant was presumed to be a prime incentive to in-migrating workers.
- C-b would master lease up to 200 living units in Meeker and Rifle. This program would presumably induce workers to utilize facilities in these two communities at an early stage of construction. Through this program C-b would convert the use of facilities in a timely manner to accommodate the permanent operations workforce.
- A Transitional Housing Program as a part of the Master Lease Program would provide operations workers with specific temporary housing upon employment. A temporary housing unit would be available to an in-migrating operations employee

for up to twelve months allowing time for securing preferable permanent housing.

- The development of an 850 temporary single-status housing facility (mancamp) at the plant site. This facility would contain a combination of recreational vehicle spaces and motel-style living quarters, which provide temporary housing for the short-term construction workers and transitional housing for operations employees.
- Expansion of permanent housing stock primarily in Meeker and Rifle by the private market place to meet the increased demand generated by the C-b operations and induced workforces. The number of living units needed between 1985 and 1989 is expected to reach 800.
- An Employee Housing Information Program. This program would provide in-migrating workers with readily accessible housing sales and rental information for the study area.
- A Housing Monitoring Program to assess changes in existing housing conditions and responsiveness of various housing mitigation measures.
- An Employee Mortgage Assistance Program, as needed, for permanent operations employees purchasing permanent housing.

These impressive demands were based upon projections that we had made for Cathedral Bluffs direct employment which started with about 500 people total employed in 1984. Construction forces peaked at about 1900 in 1986 and declined to zero by 1989. Operations workforce leveled out at 1000 people by 1989. The maximum number of people directly employed by C-b was 2500 in 1986.

These workforce estimates were then translated into housing demands for the two impacted counties using agreed-upon parameters in the aforementioned model.

The required number of housing units, which ranged from 4300 in 1984 up to 6500 in 1986 and stabilized at 5400 in 1989 is less than the number of employees due to the number of local residents who would take jobs plus occupancy of some units by two jobholders. The total required residential units included a baseline value of established units of about 3900, used by the model for all non-C-b Project-related activity within the study area. The C-b Project-related demand was described by three household classifications: direct construction, direct operations, and induced.

The direct construction workforce households represented those workers who would in-migrate to the study area to work temporarily on the construction phase of the C-b Project. The projected direct construction demand reached an annual average peak of 1,543 units in 1986, then rapidly declined to zero by the end of 1988. The major temporary construction housing demand was projected to take place between 1985 and 1987, with 1984 and 1989 as mobilization and demobilization years, respectively.

The direct operations workforce would be those employees hired directly to operate the facilities. The operations housing demand was projected to slowly increase over the project's six-year period until the demand reached an annual average peak of 779 units in 1989. The 779 units represented the project's calculated operation housing demand steady-state level.

The induced workforce refers to those jobs which would result from the overall economic growth and diversification created by the development of the C-b Project. The model predicted a level of about 800 residential units required for this population.

Again, the results of this study translated into about 2500 newly constructed units required at the peak of construction in 1986 and a permanent addition of about 1500 units for the foreseeable future. No wonder, local folks had dollar signs in their eyes. Many locals had been waiting all their life to cash in on the oil shale reserves. This appeared for a time to be their chance.

# CHAPTER 15

# The Future

## Dependent or Not?

WHAT, IF ANYTHING, could be done to foster an oil shale industry in the U.S.? There are several things that could happen. There seems to be no clear-cut choice, in my opinion. Things could go along as they are now: modest (by comparison) experimental work such as that Shell Oil is carrying on in the field, incidental information gathering by existing soda production enterprises and natural gas drilling in the Piceance Basin, and probably additional research efforts by Shell and other companies such as ExxonMobil. On the other end of the continuum, a full-blown energy crisis could develop again resulting from the instability of our world because of terrorist actions with ties to Middle East oil producing countries. All of the many arguments made in the 70's about undue reliance on off-shore energy sources could again rear their unsettling heads. Whether we do anything or not, as always, depends on the unpredictable future. I sometimes wish that our leaders would call upon Mel Brooks for help in knowing what will happen in the future. Any one who could, in 1974, so clearly see 20 years into the future (see his forecast for executive deportment as Governor William J. Le Petomane in *Blazing Saddles*) should be called upon to help us decide if, when, and how we should spend our national income on developing domestic energy sources.

Do we USers really want to be energy self-sufficient? That would not appear to be the case viewing the history of the past 20 years. If anything, our reliance on off-shore oil has increased during that time. If it really happens that sentiment develops towards increasing domestic production of petroleum to reduce our dependence on others, for national defense or other reasons, then different paths could be followed to develop the proven oil shale resources. The main choice, in my opinion, would be whether or not to develop the industry with federal government funding or to allow private industry to try to develop a competitive product using market dynamics of profit and loss. As usually exists in this sort of dilemma, there is probably some sort of middle ground that would best promote the successful development of this resource.

Let's examine some of the pros and cons of future oil shale resource development in light of lessons learned from the most recent attempt to give birth to this new industry in the western United States.

Arguing for development of an oil shale production field may be such things as:

- The above mentioned, and often repeated arguments, about the need for domestic energy self-sufficiency. This has been an increasingly popular subject after September 11, 2001 terrorist attacks. The popular press mentions more and more often that this war might include direct attacks on Iraq, which currently satisfies a significant portion of the US's daily petroleum habit. Immediate loss of that supply, plus other sympathetic OPEC supply stoppages could lead to shortages and gasoline lines. The current arguments center mostly on opening a small section of a federal wildlife reserve on the

North Slope of Alaska, but the same arguments apply to the Piceance Basin.

• An opportunity to capitalize on the many millions of dollars spent by industry in the 1970's and 80's to prove out various technologies and equally important, to prove that development could be accomplished within stringent restrictions on environmental and socio-economic impacts. If it is assumed that domestic self-sufficiency of petroleum supply is a worthy goal for our country as discussed in the previous paragraph, there is a good argument for picking up oil shale development where it left off in the 80's, simply because of all the important environmental and socio-economic background information and experience that were produced and exist in government and industry files.

• Acceptance of the Hubbert curve and realization that petroleum resources will run out and the only real question being when this will occur. The Department of Energy's Energy Information Administration, in its Annual Energy Outlook 2002, forecasts that

"World oil demand is projected to increase from 76.0 million barrels per day in 2000 to 118.9 million barrels per day in 2020 . . . due to higher projected demand in the United States and developing countries, including the Pacific Rim and Central and South America . . . . OPEC oil production is expected to reach 57.5 million barrels per day in 2020, nearly double the 30.9 million barrels per day produced in 2000, assuming sufficient capital to expand production capacity. Non-OPEC oil production is expected to increase from 45.7 to 61.1 million barrels per day between 2000 and 2020."

The report identifies non-OPEC increases in the Caspian Basin, offshore West Africa, Mexico and Brazil. The report states, perhaps ominously in agreement with Deffeyes' Hubbert Curve prediction for the whole world

"North Sea production is expected to peak in the middle of the current decade, reaching 7.5 million barrels per day."

Arguments against development could include:

- Residual bitterness against industry for seemingly thoughtless treatment of people and the area because of precipitous withdrawal from on-going projects contributing to a historical boom-bust scenario which is probably no longer acceptable to Westerners.
- Remaining uncertainties about technical questions concerning existing extraction techniques and processes.
- Appearance of possible other new sources of domestic energy supplies such as marginal natural gas fields, as the price of energy increases.
- Remaining uncertainties about overall impact on environmental and human resources.

## Dollars for Minerals

An excellent article in Shale Country by Anne Wasko, just after the Exxon pullout aptly summarized the situation then (1982) as well as now.[xiii]

"Strictly defined, according to the "American Heritage Dictionary," economics is the "science of the production, distribution and consumption of commodities." By examining factors that influence supply and demand for a product, employment, prices and so on, an economist can explain present events based on past observations, and can estimate what *might* happen in the future.

An economic analysis can be made of a large market, such as the United States, or of a single market or economic unit, such as minerals. Few people in shale country have escaped being affected by mineral economics at one time or another. For example, just two years ago the shale industry seemed on the verge of getting off the ground floor—once again. But by the end of 1981, several projects began slowing activities, some citing economics as the reason. And in May, Exxon Corporation announced it was discontinuing funding of (the) present Colony Shale Oil Project near Parachute, CO.

Dr. Charles W. Berry, head of the mineral economics department at the Colorado School of Mines in Golden, CO, emphasizes that mineral economics are so complex because more is involved than simply examining supply and demand for a commodity. Instead, he says, unlike some other types of economic analysis, "We take the process a step further by applying standard management principles to the study of minerals. Efficiency and cost of production, for example, are significant factors in the overall analysis of any minerals operation.'

In many ways, Berry notes, oil shale development is a particularly good example of how the economics of minerals differs from that of other products. 'Minerals development is quite different from most industrial situations,' he continues. 'For instance, when someone invests in a factory, the money put into the operation can be used for more than one purpose. If the factory was originally set up to manufacture a particular product and, as changes occur in consumer demand, the product is no longer wanted, the factory might be retooled to manufacture a different product.

'On the other hand,' he adds, 'when an investor puts money into a mining project, the facilities can only be used for one thing: to extract that mineral. So, when the demand for the mineral drops, the mine is unable to fulfill a different economic need—the money put into mineral development is a single-purpose investment.

'Oil shale development, like most mineral operations, is very capital intensive. A huge investment is required up front before any production can get started. In the early days of mining, the situation was just the opposite because human labor was primarily used to produce the mineral. This made it easier to start up an operation and to adjust production.' Today, he notes, equipment and construction costs have to be paid before anyone can start to extract a commodity. Thus business wants to be assured of a market for the commodity before any large investment is made.

Another problem is accurately estimating how much money will be needed for front-end production. 'Making accurate cost estimates for start-up is very difficult,' says Berry. 'For example, in the case of oil shale, I think capital costs were going up at a rapid rate, and some of the projects couldn't justify the capital expenditure because the rate of return on their investment would be inadequate. All investors, no matter what the investment, look at alternate uses for money, and invest where the rate of return is the best.'

Although the cost of production is an important aspect of whether or not a mineral is developed, the situation in other parts of the world also plays a role. Berry explains that 'With most mineral commodities you have an international market. Therefore, if the price of imported copper is less than domestic copper, manufacturers will buy from abroad rather than pay a higher price for a domestic product. As far as oil shale is concerned, when the price of petroleum dictates that it's less expensive to import crude from Venezuela, Nigeria or the Middle East than to produce a synthetic crude from oil shale, chances are that there will be little interest in pursuing synfuel development.

'And,' he continues, 'conservation has had an impact on the decision to produce oil shale. Because the country is in a recession, people have cut back on energy usage. We're using less oil and, therefore, a surplus of conventional crude exists. Basically, this means that the market is not favorable for the higher-priced synthetic shale oil.'

But, the future of oil shale is not entirely at the mercy of outside factors. By applying management principles to the economic analysis of shale, Berry notes that there are several ways the economic environment could become more favorable for shale development. One possibility, while not always feasible, would be finding ways of using shale byproducts to reduce the risk of the initial investment. Scotland, for example, had a commercial shale industry—although it was only 3,000 barrels a day—from 1850-1964. The oil's co-product was ammonia sulfate, which was used for fertilizer.

'Another possibility,' Berry notes, 'one that is most important, is to develop more effective technology. A major technological breakthrough could make it less expensive to produce oil shale and thus more cost competitive with regular petroleum.'

Even if prices for petroleum rose drastically, Berry thinks it is unlikely that this fact alone would spur a major oil shale push, because, 'The risk still exists that prices might turn down again. If that happened, developers would be in the same situation they are now, generating a product that is priced above what the market will bear.'

Since events do not occur in a vacuum, the impacts of mineral economics on the oil shale industry have in turn impacted the economics of Western Colorado and Eastern Utah. As any industry experiences rises and falls, the waves can be felt in surrounding areas. In general, the size of the impact depends on how large a percentage of the work force is involved in that industry.

Steve Campbell, an economist and senior research associate for the National Institute for Socioeconomic Research, Boulder, CO, says, 'For example, Garfield County, CO, experienced a large change when Colony shut down, because 2,100 of a total of 11,000 nonagricultural workers in the county were affected. However, most of the people laid off were temporary construction workers. Many of them were not permanent residents of the county, and therefore, they did not spend a large percentage of their income there. With this in mind, many service businesses in Garfield had not geared up for the full economic potential of the situation. So, the Colony situation did not devastate the area.

'In fact,' Campbell notes, 'as we analyzed the effects of the Colony withdrawal, we noticed that there wasn't a mass exodus from the area. Union Oil Co. of California announced it would try to hire some Colony workers for its project near Parachute, so some people who were laid off applied for those positions. Also, a number of people with families waited to finish out the school year and decided to stay.

'These people like the area, and many feel that chances for finding employment are not any better elsewhere, since the entire nation is experiencing unemployment. Now if shale development picks up or if another industry blossoms on the Western Slope, an adequate number of people are available to help build a work force.'

Campbell believes that the long-term fluctuations in the shale industry probably won't permanently affect the area's economy. 'If the ski industry expands, that will take some pressure off mining to produce as many jobs as it has in the past. Also, Union is still going ahead with its plans, and several other projects, such as the Chevron project near DeBeque and Mobil's venture near Parachute, are in the preconstruction phases.

'It's hard to make generalities about the way the rises and falls of the shale industry affect Colorado and Utah,' he continues. 'While Garfield County may be hard hit right now; Grand Junction is relatively less affected because it has a larger labor force and a more diverse economy. If the status of White River (federal tracts U-a and U-b) shifted drastically, Uintah County in Utah would feel the impact more heavily than Mesa and Garfield counties in Colorado have, because Uintah has a smaller population—the percentage of people affected would be much greater.'

Obviously, it is difficult to say precisely what economic changes lie in store for the oil shale industry or for the communities influenced by shale development. Many factors ranging from the number of employees, to the cost of transportation to events in the Middle East, play a part in shale economics. While the study of economics can provide a clue about what might happen, when speaking of shale, the term "simple economics" is truly a figure of speech, not a fact of life.'

As a brief aside here, I ran across some interesting statistics in my research which supports the energy company contention that oil is one of the best bargains around. In Welles' book noted in Chapter 1, he lists prices of petroleum products as of July 1, 1920, including gasoline at $0.20/gal and motor oil at $0.40/gal. Other products were in the same range. If we divide today's gasoline price of, say $1.50/gal by the above 20 cents, we get an increase ratio of 7.5. If we divide the year 2001's published Construction Cost Index of 6342 by the 1920 index of 251, we get an increase ratio of about 25. So, while one of the most widely cost of inflation parameters has increased 25 times, the price of gasoline has increased only 7.5 times since 1920.

## Role of Government

A related question about the whole oil shale development question concerns the appropriate role of government, particularly the federal government, in the proceedings. Most of the resource exists on federal land. The impacts, however, will occur primarily to local people and local governments. There is a never-ending argument about the best way to resolve economic problems, relying on either Adam Smith's 'invisible hand" or government bureaucratic planning. Far be it from me to attempt to decide this age-old dilemma. However, I can describe my prejudices with a local anecdotal example. It has been historic here in the Grand Valley to complain about high gasoline prices. The local newspaper could count on seasonal letters describing experiences of natives traveling in any direction from Grand Junction and finding lower gasoline prices. There was always talk of collusion and demands for investigation of monopolistic practices by regional oil companies. Indeed, there were some such investigations, which always seemed to end inconclusively. The investigations at least would help to stifle the complaints for awhile. The conclusions generally centered around the high cost of transportation of fuels to the Grand Valley from distant refineries. Interestingly enough, there was a determined, and successful, effort to shut down the only refinery in the valley because of environmental concerns. It's always hard to reconcile the two clashing goals of quality of life and cost of living no matter where one lives. In any event, the gasoline price issue has disappeared in Grand Junction simply because Sam's Club came to town and opened a gasoline station with prices considerably lower than the existing stations. Other large stores have duplicated Sam's stations and, as a result, all the prices have dropped. So much so, in fact, that recent newspaper articles talk of our prices at times being among the lowest in the state. This would seem to me to be a strong argument for allowing free enterprise and market forces to guide the development of energy supply sources.

My personal prejudices about government participation in private matters can be illustrated by thoughts I have regarding both flush toilets and filtered cigarettes. The flush toilet issue has been often discussed by columnist-comedian Dave Barry. The issue centers around mandates by well-meaning bureaucrats concerning the volume of water to be allowed per flush, ostensibly to save water. These people who apparently believe that they know what's best for others have probably actually increased the overall use of water because toilets need to be flushed multiple times to accomplish what they're supposed to do.

Cigarette filters have been less discussed. However, similar logic prevails. Again well-intentioned folks created a crusade to put cellulose filters on cigarettes in an effort to reduce the amount of nicotine and tars that smokers inhaled. The only problem is that smokers' systems depend on a certain level of nicotine to feed their addiction. So they simply smoke more filter cigarettes to maintain the same amount of nicotine in their systems. In addition, the disposal of cellulose is much more difficult that non-filtered cigarettes. I imagine the filter material is biodegradable, but certainly at a much slower rate than the plain paper and tobacco cigarette butt. This enlightenment came to me as I cleaned a section of our city streets as part of the effort to keep highways and streets cleaner. The number of cigarette filters I picked up was astounding, clearly the most pervasive waste product on the street side. So, the shortsightedness of the busy-bodies did not really reduce the amount of tobacco components inhaled per smoker and at the same time created a new, significant pollution source. In the meantime, I'm sure it made some folks who provide filter materials rich.

Local government in Western Colorado is generally realistic and guided by predominantly economic concerns. Typical local thinking is represented by a quote from one of our very important and very nice local citizens, the late Jane Quimby, at the time a Grand Junction city council woman. In a January 1979 Grand Junction Daily Sentinel article, she said, "More and more, there is the feeling that it's better not to get involved with federal or even state funding help (for community projects), if possible, because of all the strings attached." There are, however, a fair percentage of green-leaning folks who apparently believe that humans are less important than other species. This political feeling is generally reinforced more by national and statewide politicians and organizations than the locals. However, I get exasperated by the number of folks around here that like to think that they know best for all of us. A noted case in point is the construction of a traffic circle at a nearby intersection, which for years had passed cars through a four-way stop system with little trouble or public concern. It turned out the traffic circle, which cost us taxpayers over a half million dollars, was probably built so that some of our city folks could travel to Japan and receive an award for beautification of the entrance to our town. In my opinion, the traffic situation is worse. People younger than about 45 go through the circle without slowing down. We older folks all stop and wait too long for some sign that it is okay to proceed. If there haven't been more accidents with the circle than with the old sign system, I'd be surprised.

As long as I'm venting, I'd like to talk about speed limits. The issue ties in with energy in that overall average vehicle speed is proportional to fuel consumption. We have created a strange system. Speed limits are posted on all public roads and highways. However, they are not consistently enforced. I am constantly frustrated by others passing me as I cruise along at the posted limit. Indeed, I have contrived a bumper sticker FOSL that stands for Fogey Obeying Speed Limit for us senior citizens being regularly overtaken, often with the overtaker's displeasure being signaled to us.

I attribute this phenomenon to the Los Angeles freeways. In order to exist there, it is necessary to close any gaps between you and the car in front of you, irrespective of the speed. Indeed, tickets are more likely to be issued for not keeping up with the established traffic flow than for speeding. This procedure has never been fully accepted here in small-town Colorado. We FOSLs continue to blithely travel along at the posted speed limits, too often in the inside lane.

However, there is an unwritten allowance administered by police and highway patrol. It seems like it may be 5 to 10mph above the posted

numbers, depending on specifics. A case in point is the current practice
of threatening to double fines in certain zones, e.g., construction and
school zones. There is either a speed limit or there isn't. It provides the
law enforcement people with an unfair advantage. They can—or not—
stop and ticket people at their personal discretion. I would like that same
flexibility given to the county clerk. When I go in to pay my property
taxes, they could say, "Oh, that's okay Mr. Loucks." "We'll let you off with
a warning this year. But you be sure and come in next year and pay your
property taxes." C'mon, let's be realistic! My old home state, Montana,
showed its colors, as diluted by California immigrants, recently when
they reinstated a state-wide speed limit after law enforcement officers
complained that they lost all their apparent power when there was no
limit. To my mind, the old Montana restriction of "reasonable and prudent"
was more workable than the current subjective, unwritten and highly
variable system. In addition, I always favored Montana's restriction on
large trucks even when there were no limits on automobiles.

Government's role in energy production and supply has a checkered
past, however, as has been addressed in passing earlier in this brief
narrative. Political considerations enter into nearly every decision made
by major oil companies as they plan, and then, execute their plans.
Basic conditions such as tax implications, population dynamics,
environmental regulations, etc., must be factored into every decision,
large or small, of energy producers. Not only all the inter-related
particulars discussed above necessary for proper implementation of
the C-b project to comply with the lease stipulations, but day to day
decisions taken by coal miners, electrical power plants and similar
energy producers have to consider all existing and potential social-
political ramifications. The penalty for failure is extremely serious—
up to and including corporate demise.

## Reduce Scope of Development

One thought I've had for years is that a limited oil shale industry
could be developed to provide feedstocks for petrochemicals. My
understanding is that shale oil would be an acceptable base for ethylene

and propylene. The number of compounds that derive from these basic hydrocarbons is incredible. An abbreviated list of things we use every day that are based on ethylene and propylene includes: high and low density polyethylene plastic compounds; polyvinyl chloride plastics; polypropylene plastics, plus numerous solvents. For an illuminating educational experience, check the labels of your clothing and all the containers in your kitchen. The number of times that the above named plastic compounds comes up will probably surprise you, that is, if you can decipher the recycle codes on the packages of 1,2, etc. (Also, in order to read what it's made of, you may have to fight extremely hard to remove some of the currently used packaging. In the first place there are too many layers of wrapping, and in the second place, the shrink-wrapped plastic coverings require a very sharp knife, a strong arm and determination for removal.)

Given the probable restrictions of water and socio-economic infrastructure for a very large shale oil industry, a smaller production target with appropriate transportation and production facilities dedicated to petrochemicals may be practical. Statistics show that from 5 to 10% of our petroleum is used for petrochemicals. So an oil shale industry of 1 to 2 million barrels per day could be useful. I don't see any diminished demand for these materials in the world. Plus it is important to realize that at this time we have no substitutes available for petroleum as a petrochemical feedstock. So, if petroleum were to be restricted and used solely for transportation, our modern life would be greatly impacted by the loss of these synthetic materials. Perhaps a developing oil shale industry could supply a critical component of consuming society's needs.

Another possibility for a limited and targeted oil shale industry could be the military. The following was delivered to Stanford Research Institute by Secretary of the Air Force John C. Stetson in 1979.[xiv]

"The United States Air Force feels very strongly about the need for a domestic synthetic fuels industry and I welcome this chance to take our case to a group such as yours. It is encouraging to see an organization with the experience and reputation of SRI—

and the backing of industry—taking on a project as important as the Synthetics Study.

Perhaps some of you are curious as to why the Air Force has such an interest in energy research. The answer is quite simple.

The Air Force has a strategic planning responsibility for the operation of our total weapon systems into the 1990's and beyond. And that responsibility includes making certain that we have materials, fuels and supporting logistics. We have no business, for example, planning and buying aircraft using titanium or unique aluminum alloys if we cannot be confident that those materials will be available for the next 20 years. Even more important, we should not be building aircraft and weapon inventories based on the use of hydrocarbon liquid fuels, unless we can be certain, that such fuels will be readily available to us.

Unfortunately, in driving airplanes we have no alternative for hydrocarbon fuels. Nuclear power plants have been suggested, but I assure you that their cost would be enormous even if they were technically feasible.

So, we have no effective alternatives. But, one then asks, is the problem really big enough to take our time and management resources. You be the judge after I give you some facts.

Today, the Defense Department is the country's largest single user of energy. It uses about 2.5 percent of the nation's petroleum consumption through direct usage. The Air Force uses one-half of that fuel for turbine engines—about 232,000 (1978 figures) barrels a day.

If the Air Force had to rely on secure domestic sources only, the figure would be eight percent of the crude produced in the United States. This assumes an optimum yield of 40 percent JP-4 per barrel. This could easily go up to 20 percent or more in wartime. If you add Navy requirements, Army requirements and the entire range of other critical defense requirements you can quickly envision what a shortage in supply could really do to us. And, this doesn't take into account the minimum essential requirements of the industrial and commercial economy, necessary to support the civilian population during war.

Another and integral part of our fuel problem is cost. If intense domestic and international competition continues to push up the cost, we will be forced to divert funds from critical manpower and weapon needs simply to take care of our fuel requirements. Such a course would have a serious impact on our military forces. These facts suggest that we have a legitimate reason for concern.

I might add, the *worst* of the problem will not come with conflict itself. During a war we could rely on a war reserve of jet fuel which, hopefully, we will have built-up beforehand. The Air Force will get what it needs to pursue the war.

The more serious problem is the many years of peace between wars. If the Defense Department has to compete with civilian allocations and restrictions on the use of fuel, I can assure you, that when there are long lines of automobiles at the gasoline station, the Air Force will be one of the last in line—and those effects will be felt heavily in our combat readiness.

I should explain that even during recent years of relative fuel availability Air Force flying hours for training have been reduced. Average flying hours per month for pilots are down by as much as 30 percent for some weapon systems compared to pre-1973 oil embargo flying rates.

This is to the point where we are losing experienced and valuable pilots because they're not flying enough. We cannot lose these pilots and still have a viable Air Force. In any major war we all will depend largely on the Air Force to fight that war in the initial stages. And, if we *lose* in the initial stages, then we don't have to worry much about the later stages.

My reason for bringing this all up is not to elicit sympathy for the problems of the Air Force, because as things go, the Air Force is doing very well. It's the trends which bother me and the conviction that unless we begin to lick our fundamental liquid fuels problem and the larger energy problem this country inevitably will grow weaker from a military standpoint.

So what can the Air Force and the Department of Defense do about it? Well, we can come up with some potential solutions or even partial ones and then promote them with anyone who will listen. We also can take steps as a major consumer to prepare for potential new fuel alternatives. And, these are the things we are trying to do now.

We believe the best solution, or partial solution if you will, to our fuels problem is synthetic fuels, manufactured from resources within the United States. The most obvious of these being coal and oil shale. We know that either one can be turned into acceptable jet fuel, but we are attracted to oil shale because of the hydrocarbon chemical structure of shale (as) compared to coal, and because our studies show it will be cheaper and better than a coal-derived fuel for turbine engines.

Of equal significance is the fact that there is perhaps as much as two trillion barrels of shale oil in just four of our Western states. This is several times more than the natural crude reserves of the Middle East. Even if this country could extract only a fraction of these resources, it would make an important contribution to our liquid fuel supply in the last ten years of this century.

We in the Air Force also are aggressively developing shale oil fuel specifications and conducting refining and engine performance cost studies. As a consumer, we need to be prepared to accept new products as they become available, and we intend to be prepared.

The basic premise underlying the Department of Defense's current synthetic fuel technology programs is that ultimate large scale commercialization of synthetic fuels will be sponsored by industry, in cooperation with the Department of Energy. We anticipate delivery of worthwhile quantities of synthetic fuels to defense perhaps as early as the mid-1980's. We recognize this may be an ambitious forecast, but the Air Force is in a unique position to help make it happen. The Air Force, because it is the single largest consumer of petroleum in the world, can provide a well-defined market for synthetic fuel produced under demonstration or commercialization programs. We are capable of assessing the long term impact synthetic fuels might have on life cycle costs of aircraft systems. Thus, we can fulfill the roles of an incentivizer, a customer, and an interested assessor of programs which would lead to the expansion of this country's synthetic fuels resources. And, we are eager to take on this job.

Before you get the idea that I may be getting too far out on a limb, I should tell you we are well aware of the many things working against shale oil: the environmental problems, the regulatory issues, and other limiting factors. But from a security standpoint, this country has no good alternative, but to overcome these obstacles.

It might be helpful to talk a little about the problems we face from various groups who are opposed to, or skeptical of the development of the nation's shale resources. The objections probably start with those that are most emotional. They involve the esthetic environment and the perceived health hazards. With regard to esthetics, no one can argue that a shale oil project won't change a local landscape. It will. But then our landscape is being changed everyday and this country has sufficient knowledge of land reformation so that even with projects of this scale we can minimize and substantially localize the esthetic and environmental impact. If our forefathers had opposed change, for the sake of change itself, much of our industrial strength would never have been built and this country might very well be a third world, non-industrialized, nation today. I don't think the American people are ready to go back to cabins and washboards as a way of life.

On the issue of health hazards—much or most of what we all do constitutes some kind of a hazard; mining coal, drilling for oil, driving automobiles, using electricity in our homes, even walking down the street. We accept these risks because they are part of the price and pleasure of living.

One final broad and all-encompassing objection is this. Shale oil will cost too much. Of course, I do not know yet, just how much is, too much. I feel certain though that our shale reserves are a source that could be depended on. And, higher cost may become a moot point if natural crude supply is unavailable.

I have heard responsible people in the oil industry state that they are willing to make substantial investments in shale oil recovery with an incentive as low as $3/barrel tax credit and a price on the oil equivalent to OPEC pricing whatever that may now be. My personal view is that a modest premium is justified in exchange for the possible and even likely development of a major energy resource. We need shale oil recovery badly, and as every month goes by this fact should become more and more evident to everyone.

I do not know how many Persian Gulf crises and Venezuela price increases it will take to drive home the point to the American people. We must make some sacrifices on consumer goods today if we are to have a more secure economic environment tomorrow. And, shale oil can very well make a material contribution to that economic environment as well as to our military strength."

## Revitalize Nuclear Power

A related issue is the possible rejuvenation of nuclear power. This system of generating power has had a checkered past and an uncertain future. From a hard engineer's viewpoint, nuclear generation has apparent advantages. As stated by Peter Huber,

"Hard power extracts more power from less of the Earth's surface. Uranium is harder than oil and gas, which are harder than coal, which is harder than biomass, solar, and wind. The greenest fuels are the ones that contain the most energy per acre of land covered, cultivated, paved, or stripped. Per unit of power produced, softer fuels consume more material, labor and—above all, land . . . .

Per unit of output, large centralized, industrial plants are usually much cleaner than the decentralized, low-tech alternatives they displace. It is far more efficient to burn oil in the huge, well-maintained boiler of a central power plant than to burn it in a two-stroke engine of a lawnmower, even after we allow for all additional losses in transmitting electricity from the power plant to the end user. It is more efficient, and cleaner, to burn fuel and distribute electricity than to refine fuel and distribute gas or gasoline.

The technology of atom and photon is the greenest of all. Nuclear power, the original "solar," extracts limitless energy from the tiniest amounts of material because it extracts subatomically."[xv]

Nuclear power generation costs have been becoming more competitive according to some. (e.g., "Nuclear Power in a Restructured Market". Nuclear News, August 2000). Costs would be even more competitive if market forces were better able to factor in the true cost of environmental pollution. If another of Huber's proposals, the privatizing of pollution were implemented, as he puts it:

"Serious pollutants are best established in ways that neither expand the public sphere nor undermine established private rights. Issuing permits in quantities that mirror established patterns of activity and use is usually the best and most economically efficient way to proceed. Going forward, the market creates direct competitive substitutions for pollution permits—pollution abatement technology. The more freely people can buy, sell, and trade pollution and permits, the more pollution will abate."

All of the above must be placed in the perspective of what has happened in the US and the rest of the world as the nuclear power industry has grown. Presidents have disagreed on strategy for the proper utilization of nuclear power and the topic has become predominantly political. The above words by Huber represent one end of the continuum. Complete abolishment of nuclear generators is the other end. President Nixon believed that nuclear power generation, with an attendant faith in technology, would be a main element of the US's recovery from the 1970's energy crisis. President Carter, on the other hand, seemed to view nuclear power as a last resort, only to be considered if conservation efforts fell short. During this time, the movie, *The China Syndrome* was released which colorfully portrayed a nuclear disaster. At the same time, Pennsylvania's Three Mile Island reactor suffered a serious emergency. Public confidence in nuclear power dropped dramatically along with political support. After a period of lots of talk and little concrete action, Congress mandated that nuclear power plants be made safer and located in remote areas. The costs of building nuclear power plants escalated at a rate even more than the extraordinary inflation rate that the country was experiencing at the time. The cost escalation was greatly exacerbated by the long delays in construction forced by lawsuits over uncertainties in regulations, helped by timid bureaucrats and aggressive environmentalists.

## Back to the Iron Horse

Given that a long term, hopefully smooth transition to a political-economic system that relies less on petroleum is necessary, a step backwards in time to a better utilized railroad system could help. Increased use of railroads for mass transit and freight haulage rather than over-the-highway cars, buses and trucks would smooth the transition by increasing the number of passenger miles and tons-of-freight miles per gallon of fuel consumed. The history of our rail system has been a sad demonstration of human greed by short-sighted capitalists and labor leaders. A realistic negotiation could result in a profitable and private railroad system.

## Water—Western US Life Blood

A particular concern of mine in the debate about establishment of an oil shale industry is the use of water. As has been stated, the consumptive use of water for known shale oil extraction technologies is, in round numbers, at least a barrel of water for each barrel of oil produced. Until I lived in this country, I might have shrugged this statistic off as "just another technological problem that we engineers can take care of." The problem is much more serious than that and one that continuously commands prime attention of a significant number of people in Colorado as well as areas downstream, e.g., Utah, Arizona, Nevada, California and Mexico. Arguments concerning the use and allocation of water are continuous, vociferous and costly. It starts locally with disagreements about shipping Colorado River Basin water to the eastern side of the Rockies to allow Denver and surrounding areas to grow. In early 2002, an attempt to reapportion the state's congressional districts as a result of the 2000 census was held up by, among other things, the perception by some that certain gerrymandering would soon result in more western water heading east. The federal government has been trying for years to force California; in particular, along with Arizona and Nevada to stop using water that has never been allocated to them. Californians, especially agriculturists, protest vehemently about losing their opportunity to make a living and, in the process, impacting the nation's ability to feed the world's hungry folk.

To dramatically demonstrate the seriousness of the situation, a seemingly innocent statement in the Exxon "White Paper" will be repeated as follows: "**It appears financially and physically feasible to bring the additional needed water to the Piceance and Powder River Basins from the Missouri and several other Western rivers—at a resulting increase in costs of about 1 percent. As a side benefit, the transportation systems also could provide water for agricultural, municipal and other uses.**" (my emphasis added, of course.) I can not imagine people sitting still while we installed a pipeline or canal from somewhere in Montana or the Dakotas down to Colorado. I have always wondered about Exxon's contention concerning increased costs and their basis for coming up with one percent of something.

## Effect of Media and Attorneys

Let me express my feelings about the effect of the journalism and legal professions on our shale effort and on major projects in general. In a nutshell, my feelings are nicely expressed by the following fictional conversation, as depicted by Howard Bahr, of two Civil War veterans heading home after the battles :

"I'll tell you, pard—maybe I ought to take up another line of work."

"Well, what would you do? Be a lawyer again?"

"No, no—I am a Christian now, since the war."

"Well, you could start a newspaper again. No doubt Cumberland
        will need one. You could—"

"No," said Stribling. "I am hampered by principles now, so that
        avenue is closed as well."[xvi]

Many of my concerns about the press center about what I always
see as the overuse of negativity as a means of selling more copies,
screen time, etc. An old speech from my files by Professor John McKetta
of the University of Texas illustrates this in a creative way.

"On March 27, 1973, I heard Garner Ted Armstrong say
over the television: "There is no way you can have any optimism
for the continuation of life on this earth because of the pollution,
over population, and results of technological advances."

It bothers me that there are so many purveyors of gloom who
talk about the hopelessness of our future.

There is an entire spectrum, from zero to Infinity, of views
and actions on almost any problem. Let's take the pollution
problem, for example. We all know there are still some companies
and cities who put toxic gases and liquids into our air and streams.
It's almost unbelievable that many of our large cities still discharge
raw sewage, or only partially treated sewage into our streams. Both
industry and the cities should be stopped immediately from these
flagrant violations. On the other extreme we have those people
who wish to have distilled water in the streams and zero particulates
in the atmosphere. These are impossible concentrations and could
not be attained even if we had no people on this earth. The answer,
obviously, is somewhere between these two extremes.

We're all deeply concerned about reports of the destruction of
our environment as a result of technological recklessness, over-
population, and the lack of consideration to the preservation of
nature. As Chairman of the National Air Quality Management
Committee I have to read great amounts of technical literature in

this area. I've turned up a lot of information that I'd like to share with you.

For example, you know that electricity is perhaps the most convenient form of energy available. Now, some people say the bad industry that produces electricity is an evil organization of the establishment whose objective is to create new radiation hazards with nuclear power plants, cut down trees, stick poles into the ground and pump smoke into the air to poison all of us.

It's a gloomy picture indeed. But I've found out that in many, many cases this outlook is not justified. This is what I'd like to talk to you about. I'm speaking to you as one who understands elementary science and engineering and not as a spokesman for the environment or ecology. Some of the facts I will mention will surprise many of you. My statements are supported by evidence that is difficult to interpret in any other way. I'll leave a list of literature citations with the chairman of your group. Read the articles to get your own interpretations. (Note: this list is included at the end of this citation.)

## 1. WHY IS THE OXYGEN DISAPPEARING??

My first surprise concerns the air we breathe. You have been reading that we are seriously depleting the oxygen in the atmosphere and replacing it with toxic substances such as carbon monoxide.

We've always been taught that oxygen in our atmosphere is supplied by green plants using the process of photosynthesis. We know that plants take in carbon dioxide and through activation by sunlight, combine $CO_2$ with water to make starches and cellulose, and give off oxygen. In this way the whole chain of plant and animal life is sustained by energy from the sun. When the vegetable or animal materials thus produced are eaten, burned, or allowed to decay they combine with oxygen and return to the carbon dioxide and water from whence they came. We all know this. Then, what is the surprise?

Surprise number one is that most of the oxygen in the atmosphere *doesn't* come from photosynthesis. The evidence is now overwhelming that photosynthesis is just inadequate to have produced the amount of oxygen that is present in our atmosphere. The amount of oxygen produced by photosynthesis is just exactly enough to convert the plant tissue back to the carbon dioxide and water from which it came. The net gain in oxygen due to photosynthesis is extremely small. The oxygen in the atmosphere had to come from another source. The most likely possibility involves the photo dissociation of water vapor in the upper atmosphere by high energy rays from the sun and by cosmic rays.

This means that the supply of oxygen in the atmosphere is virtually unlimited. It is not threatened by man's activities in any significant way. If all the organic material on earth were oxidized, it would reduce the atmospheric concentration of oxygen by less than 1%. We can forget the depletion of oxygen in the atmosphere and get on with the solution of more serious problems.

## 2. *CARBON MONOXIDE WILL KILL US ALL!*

As you know, the most toxic component of automobile exhaust is carbon monoxide. Each year man adds 270 million tons of carbon monoxide to the atmosphere. Most of this comes from automobiles. People are concerned about the accumulation of this toxic material because they know that it has a life in dry air of about 3 years. Monitoring stations on land and sea have been measuring the carbon monoxide content of the atmosphere. Since there are 9 times more automobiles in the Northern Hemisphere than in the Southern Hemisphere it is expected that the Northern Hemisphere will have a much higher concentration of atmospheric carbon monoxide. The true measurements show however that there is no difference in CO amounts between the hemispheres and that the overall concentration in the air is not increasing at all. In fact they've found higher concentrations of CO over the Atlantic and Pacific oceans than over land ???

Early in 1971 scientists at the Stanford Research Institute in Palo Alto disclosed that they had done some experiments in smog chambers containing soil. They reported that carbon monoxide rapidly disappeared from the chamber. After sterilizing the soil they found that now the carbon monoxide did not disappear. They quickly identified that organisms were responsible for CO disappearance. These organisms, on a world-wide basis, are using all of the 270 million tons of the CO made by man for their own metabolism, thus enriching the soils of the forest and the fields.

This does not say carbon monoxide is any less toxic. It does say that, in spite of man's activities, carbon monoxide will never build up in the atmosphere to a dangerous level except on a localized basis. To put things in perspective, let me point out that the average concentration of CO in Austin, Texas is about 1.5 parts/million. In downtown Houston, in heavy traffic, it sometimes builds up to 15 or 20 ppm. In Los Angeles it gets to be as high as 35 ppm. In parking garages and tunnels it is sometimes 50 ppm. Here lies surprise number two for you—do you know that the CO content of cigarette smoke is as high as 42,000 ppm? The CO concentration in practically any smoke filled room grossly exceeds the safety standards we allow in our laboratories. Of course, 35 to 50 ppm CO should not be ignored but there are so many of us who subject ourselves to CO concentrations voluntarily (and involuntarily) that are greater than those of our worse polluted cities, including Holland Tunnel in New York, without any catastrophic effects. It is not at all unusual for CO concentrations to reach 100-200 ppm range in poorly ventilated, smoke filled rooms. Incidentally, if a heavy smoker spends several hours without smoking in a highly polluted city air containing 35 ppm of CO concentration, the concentration of CO in his blood will actually decrease! In the broad expanse of our natural air, CO levels are totally safe for human beings. We all know that we should not start our automobiles in closed unventilated garages.

### 3.  OXIDES OF NITROGEN WILL CHOKE US!

One cannot help but be extremely impressed by the various research efforts on the part of petroleum, automotive, and chemical companies to remove oxides of nitrogen from the products of combustion in the tail pipe gas of our automobiles. You've read about the brilliant work of Dr. Haagen-Smit that showed that the oxides of nitrogen play a critical role in the chain reaction of photochemical smog formation in Los Angeles. Oxides of nitrogen are definitely problems in places where temperature inversions trap the air.

But we've all known for many years that nature also produced oxides of nitrogen. The number three surprise (and shock) is that most of the oxides of nitrogen *come from nature*. If we consider only nitric oxide and nitrogen dioxide the best estimates are 97% is natural and only 3% is man made. If we also consider nitrous oxide and amines, then it turns out that 99+% is natural and less than 1% is man made.

The significance of this is that even if we are 100% successful in our removal of the oxides of nitrogen from combustion gases, we will still have more than 99% left in the atmosphere which is produced by nature.

### 4.  THE DEATH OF LAKE ERIE

We've all read for some time that Lake Erie is dead. It's true that the beaches are no longer swim able in the Cleveland area and the oxygen content at the bottom of the lake is decreasing. This is called eutrophication. The blame has been placed on phosphates as the cause of this situation. Housewives were urged to curb the use of phosphate detergents. In fact for several years phosphate detergents were taken off the market. There's been a change in law since scientific evidence proved that the phosphate detergents were not the only culprits and never should have been removed from the market in the first place.

Some studies show clearly that the cause of the eutrophication of Lake Erie has not been properly defined. This evidence suggests that if we totally stopped using phosphate detergents it would have no effect whatever on the eutrophication of Lake Erie. Many experiments have now been carried out that bring surprise number four—that it is the organic carbon content *from sewage* that is using up the oxygen in the lake and not the phosphates in the detergents. One must be extremely careful in these studies since the most recent report by Dr. D. W. Shindler states that phosphorous is the culprit rather than the organic carbon. But the reason that the Cleveland area beaches are not swim able is that the coliform bacterial count is too high, not that there is too much detergent in the water.

Enlarged and improved sewage treatment facilities by Detroit, Toledo, Sandusky, and Cleveland will be required to correct this situation. Our garbage disposal units do far more to pollute Lake Erie than do the phosphate detergents. If we put in the proper sewage treatment facilities, the lake will sparkle blue again in a very few years.

Incidentally, we've all heard that Lake Superior is so much larger, cleaner, and nicer than Lake Erie. It's kind of strange then to learn, that in 1972 and 1973 more tons of commercial fish were taken from Lake Erie than were taken from Lake Superior.

## 5. DDT IS THE WORST THING THAT EVER HAPPENED TO US????

DDT and other chlorinated compounds are supposedly endangering the lives of mankind and eliminating some bird species by the thinning of the egg shells of birds. There is a big question mark as to whether or not this is true. Even if it is true, it's quite possible that the desirable properties of DDT so greatly outnumber the undesirable ones that it might prove to be a serious mistake to ban entirely this remarkable chemical.

Many of you heard of Dr. Norman E. Borlaug, the Nobel Peace prize winner. He is opposed to the banning of DDT.

Obviously he is a competent scientist. He won the Nobel prize because he was able to develop a new strain of wheat that can double the food production per acre anywhere in the world that it is grown.

Dr. Borlaug said "If DDT is banned by the United States, I have wasted my life's work. I have dedicated myself to finding better methods of feeding the world's starving population. Without DDT and other important agricultural chemicals, our goals are simply unattainable."

DDT has had a miraculous impact on arresting insect borne diseases and increasing grain production from fields once ravaged by insects. According to the World Health Organization, malaria fatalities alone dropped from 4 million a year in the 1930's to less than 1 million per year in 1968. Other insect borne diseases, such as encephalitis, yellow fever, and typhus fever showed similar declines. Surprise number five is that it has been estimated that 100 million human beings who would have died of these afflictions are alive today because of DDT. Incidentally recent tests indicate that the thinning of bird egg shells may have been caused by mercury compounds rather than DDT!

## 6. WE'RE KILLING OFF TOO MANY SPECIES!

Many people feel that mankind is the one responsible for the disappearance of the animal species. The abundance of evidence indicates that he has very little to do with it. About 50 species are expected to disappear during this century. It is also true that 50 species became extinct last century and 50 species the century before that and so on. Dr. T. H. Jukes of the University of California points out that about 100 million species of animal life have become extinct since life began on this planet, about 3 billion years ago. Animals come and animals disappear. This is the essence of evolution as Mr. Darwin pointed out many years ago. Mankind is a relatively recent visitor here. Surprise number six is that one of man's failures is that he has *not* been successful in eliminating a single insect species—in spite of his, all out war on certain undesirable ones in recent years.

## 7. MAN IS THE REAL POLLUTER!

Here's the *seventh surprise!* The late Dr. William Pecora reported that all of man's air pollution during his thousands of years of life on earth does *not* equal the amount of particulate and noxious gases from just three volcanoes, (Krakatoa, near Java.-1883; Mt. Katmai, Alaska-1912; Hekla, Iceland-1947)

Dr. Pecora pointed out that nature s pure water is not so pure after all. Here are a few of his examples:

1. The natural springs feeding the Arkansas and Red Rivers carry approximately 17 tons of salt *per minute.*

2. The Lemonade Springs in New Mexico carry approximately 900 pounds $H_2SO_4$ per million pounds of water. (This is more than ten times the acid concentration in coal mine discharges.)

3. The Mississippi River carries over 2 million tons of natural sediment into the Gulf of Mexico *each day.*

4. The Paria River of Arizona and Utah carries as much as 500 times more natural sediment per unit volume than the Mississippi River. The Mississippi sediment concentration ranges from 100 to 1,000 mg per liter. The Paria River concentration has been measured as high as 780,000 mg per liter during 1973.

*LET'S GO BACK TO THE GOOD OLD DAYS.* Dr. Isaac Asimov admonishes us not to believe the trash about the happy lives that people once had before all this nasty industrialization came along. There was no such thing. One of his neighbors once asked him "What has all these 2000 years of development of industry and civilization done for us? Wouldn't we have been happier in 100 B.C.?" Dr. Asimov said "No, chances are 97 out of

100 that, if you were not a poor slave, you'd be a poor farmer, living at bare subsistence level."

When people think of ancient times, they think of themselves as members of aristocracy. They are sitting in the Agora in Athens listening to Socrates in the Senate House in Rome debating with Cicero, riding on horses as knights of Charlemagn (sic) time. They are never slaves, never peasants, BUT that's what most of them would be.

My wife once said to me "If we lived a hundred years ago we'd have no trouble getting servants". I said, "If we'd lived 150 years ago we'd be the servants."

Let's consider what life was really like in America just 150 years ago. For one thing, we didn't have to worry about pollution very long—because life was very brief. Life expectancy of males was about 38 years of age. It was a grueling 38 years. The work week was 72 hours. The women's lot was even worse. They worked 98 hours a week scrubbing floors, making clothes by hand, bringing in fire wood, cooking in heavy iron pots, fighting off insects without pesticides. Most of the clothes were rags by present day standards. There were no fresh vegetables in winter. Vitamin deficiency diseases were prevalent. Homes were cold in winter and sweltering in the summer.

Epidemics were expected yearly and chances were high that they would carry off some members of the immediate family. If you think the water pollution is bad now, it was *deadly* then. In 1793 one person in every five in the city of Philadelphia died in a single epidemic of typhoid as a result of polluted water. Many people of that time never heard a symphony orchestra, or traveled more than 20 miles from their birth place during their entire life time. Many informed people do not want to return to the "paradise" of 150 years ago. Perhaps the simple life was not so simple.

*WE ARE PRACTICING WITCHCRAFT.* Dr. A. Letcher Jones points out that in every age we have people practicing witchcraft in one form or another. We all think that the people of New England were irrational in accusing certain women of being witches without evidence to prove it. Suppose someone accused you of being a witch? How could you prove you were not? It is impossible to prove unless you can give evidence. It is precisely this same witchcraft practice that is being used to deter the construction of nuclear power plants. The opponents are saying that these plants are witches and it is up to the builders and owners to prove that they are not. The scientific evidence is that the nuclear power plants, constructed to date, are the cleanest and least polluting devices for generating electricity so far developed by man. We need electricity to maintain the standard of living we have reached but to the extreme environmentalists we are witches. We should be burned at stake.

We hear the same accusations about lead compounds from the gasoline engine. Our Environmental Protection Agency has *no* evidence that there has ever been a single case of death, or even illness from lead in the air coming from burning of gasoline, but they still insist that we must remove the lead from the gasoline. To the EPA we are witches. They have no evidence—no proof—we are pronounced guilty! And yet you know that gasoline needs some additives to prevent engine knocks. If we don't use tetraethyl-lead we'll have to use aromatic compounds. Some aromatics are carcinogenic. We know that! The use of unleaded gasoline also can use up to 12% more crude oil. (Incidentally, the real reason for removing lead from gasoline was because it was suspected that lead poisoned the catalyst in the emission control unit. Now we have some evidence that it isn't the lead but ethylene bromide which is the poisoner.)

From what we read and hear it would seem that we are on the edge of impending doom. A scientific evaluation of the evidence does not support this conclusion. Of course we have many undesirable problems attributed to technological activities. The solution of these problems will require a technical understanding of their nature not through emotion. They cannot be solved unless properly identified, which will require more technically trained people—*not less.*

Thomas Jefferson said that if the public is properly informed, the people will make wise decisions. The public has not been getting all of the facts on matters relating to ecology. This is the reason why some of us are speaking out on this subject today—as technical people and as citizens.

## 8.  THE LAST SURPRISE (#8)—WE'RE GOING TO LIVE.!

In summary let me state that we are not on the brink of an ecological disaster. Our oxygen is not disappearing. There will be no build up of poisonous CO. The waters can be made pure again by adequate sewage treatment plants. The disappearance of species is natural. A large percentage of pollution is natural pollution and would be here whether or not man was on this earth. We cannot solve our real problems unless we attack them on the basis of what we know rather than what we don't know. Let us use our knowledge and *not our fears* to solve the real problems of our environment.

## Literature Cited

1.   Anon, Chem. & Eng. News, p. 24, May 10, 1971.

2.   Broecker, W. S., Man's Oxygen Reserves", Science, *168*, p. 1537, June 26, 1970.

3.   Inrnan, R. E., and Ingersoll, R. B.,"Uptake of Carbon Monoxide by—Soil Fungi", J. Air Pollution Cont. Assoc., *21*, No. 10, p. 646 (Oct. 1971).

4.   Merriman, D., "The Calefaction of a River", Scientific American, p. 42-52, May 1970.

5.    Mitchell, D., "Eutrophication of Lake Water Microcosms: Phosphate vs. Non-phosphate Detergents", Science, *174*, p. 827, Nov. 19, 1971

6.   Niesler, R. A., "Industrial Emissions: An Analysis of Some Key Factors", J.Inst. Pet. *56*, No. 552, p. 344 (1970).

7.   Ochsner, A., "Hazards of Air Pollution—Fact or Fiction?" Proc. Am. Power Conf, *31* 23 (1969).

8.  Peters, M. S., Chem. Eng. Prog.Symp. Ser., 67, #115, p. 1 (1971).
9.  Stanford Res. Inst. J. *23*, 4-8 (Dec. 1968).
10. U. S. Department of Health, HEW, "Smoke and Health", No. 1103, Chap. 6, p. 49-65, 196
11. van Valen, L., "The History and Stability of Atmospheric Oxygen", Science, *171*, p. 439, Feb. 5, 1971.
12. B. Maxwell Teague, Chief Res. Scientist, Chrysler Corp. 1974. (See Chem. Eng. News, May 13, page 5 (1974) for digest of his report.)

The majority of this speech is drawn heavily from talks given by my friends, Dr. A. Letcher Jones, Dr. I. W. Tucker and the late Dr. William Pecora."

The local newspaper and TV stations were enthusiastic and generally favorable to the development of an oil shale industry, at least until the Exxon pullout. Personally, they allowed me my allocated "15 minutes of fame" more than once by printing articles and pictures of visits by US Senators Haskell and Hart and Vice President Mondale to the oil shale area. The write-ups were generally positive to our developmental effort.

After 1982, after the impacts of the precipitous pullout from the Colony project, most press was critical. I can't say that I blame them or anyone for being critical. Colorado has experienced a number of boom and bust cycles from mineral development influenced by politics and economics. The 1982 bust was taken from the same mold: no advance warnings, a business as usual manner right up until the day before the crushing announcement, and an apparently cavalier attitude. Those of us who were still working on our projects were extremely concerned by both the impact on us and the community. There was an effort, particularly by Union Oil, to absorb some of the job losses and economic hardships imposed on the community infrastructure by the shutdown of Colony.

As far as lawyers are concerned, I personally think that considerably more progress could be made if they were only available on a retainer basis. During my many years with Shell Oil, I can remember having to deal with the legal staff only once and that during a troublesome labor dispute. To the contrary, at Oxy and C-b, we had resident attorneys who were involved in everything. I always figured that at least part of the reason for so much exposure to legal counsel was due to Dr Hammer's rather flamboyant entrepreneurial manner. He and his board of directors both probably felt more comfortable with ready legal oversight and advice available. Another reason was the fact that C-b was always a joint venture among oil companies and the respective management probably felt better if lawyers were directly involved with joint contracts and agreements. The other reason for the undue influence of lawyers can be seen by looking again at the "Letter of Intent" signed by the SFC and the C-b partners outlined in Chapter 12. The caveats and legalese are impressive if not dazzling.

Other problems I have with legally trained folks is their need to be "spin doctors" and attempt to manipulate stories to further their parochial interests. This is an automatic action resulting from their training in the adversarial nature of our legal system. Also, this same training makes it appear that lawyers would have us all live a risk free life. It is counter to my upbringing to have government, led by lawyers, attempt to make everything the fault of someone else. We need to accept responsibility for our own actions and not look for scapegoats. I notice as I write this that I'm drinking coffee from one of my favorite cups which alerts me as follows: "CAUTION Contents May be Hot." Well, I hope so.

From my viewpoint, a good example of why lawyers do not need to be an integral part of project operations, was our initial dealings with the Department of Interior during the early days of the C-b lease when Shell was the operator. As complex and potentially adversarial as the Prototype Leasing Program with its Federal Oil Shale Supervisor and the unique Oil Shale Environmental Advisory Panel was, lawyers were not part of the workings. Things went very well. So I believe that the overwhelming presence of lawyers is detrimental to the progress of project work and future oil shale proceedings should be structured to minimize their influence.

# CHAPTER 16

# Conclusions

WHAT SHOULD WE do? As always the answer is, "It depends." Given all my prejudices and my background, I would like to see an oil shale industry developed. The main reasons influencing my feelings include the following:

- So much work was done in the 1970's and 1980's to provide facts and figures necessary for moving ahead with a real industry.
- Inflation has been brought under control and appears as though it will continue to be controlled in the future.
- There is adequate reason to once again put credibility in Hubbert's curve and begin to prepare for the eventuality that world petroleum production will peak in the foreseeable future.
- International politics, especially the US war on terrorism, create increasing risks of large-scale petroleum embargos and unreasonable price manipulations.

There will be many who disagree with me for all the reasons heretofore discussed. The important question to me is, however, just how truly important to we USers are our beliefs about reducing petroleum imports with concomitant dependence on foreign nations? I would have been more relaxed in the recent past, before we were attacked in New York and Washington. Our subsequent reaction against bin-Laden and his organizations and our determination to continue the war against similar terrorists wherever they may be located is greatly increasing the risk of retaliation by major international petroleum suppliers. Bin-Laden is from Saudi Arabia, the most powerful oil nation in the world. One of our talked about targets is Iraq, which currently supplies us with significant amounts of oil. All these things make me more and more concerned about the continued reliability of our offshore oil supplies.

I believe that there is not enough conventionally produced oil left in our country to supply our needs, no matter how much more exploration we do. In other words, I've become a firm believer in Hubbert's curve and its underlying calculations. Other domestic resources are necessary, as a result. I also believe that nuclear power could safely be developed to supply all the nation's electricity needs. None of this is easy, but I do believe it can be done. It's time to act to protect our wonderful country for current and future generations.

Even when and if all of our electricity is produced from home grown nuclear fuels; there will still be a significant need for petroleum. The manufacture of petrochemicals has no other feedstock than petroleum. There is no alternative for transportation fuels other than petroleum.

So why not take advantage of all the millions of dollars we have spent on the development of oil shale and go ahead and commercialize the industry? There are more than enough data available to frame answers to any question that may arise about the effects of an oil shale operation on any part of the ecology. There are sufficient data to provide answers to any concerns that anyone may have about the impact of such an industry on the surrounding communities of people.

Although it can thus be stated that most of the pressing concerns identified in the prototype oil shale program EIS and program statements have been answered, residual concerns remain. One of the concerns that came up from time to time in the 1970-80 oil shale development was the use and availability of construction workers. Most of the accomplished work was done with open-shop constructions firms, which is readily acceptable on the western slope of Colorado. However, the utilization of Davis-Bacon rules and rates which minimizes the cost effect of open shop labor were mandated for C-b because of the use and influence of federal funds on the development. Some agreement should be worked out with organized labor before future development starts.

Also concerning labor availability was the far-sightedness of Clarke Watson, a black visionary from Denver. Clarke could see the likelihood of a large number of black workers being used in the construction and operation of the industry. At the time, he was especially concerned about a disproportionate share of blacks being unemployed and saw the oil shale business as an opportunity to raise the overall standard of living of this part of our population. The same situation may obtain today and a well thought out partnership may pay significant dividends for a budding industry.

There are not sufficient data available on the reliable operability of a shale retorting process. Any program undertaken would still have to be a prototype program, in which various processing schemes are tried, debugged and either accepted or discarded. There is, however, a tremendous amount of process information that has been generated which can eliminate some options and provide direction towards those most likely to succeed.

I believe that both Mssrs. Welles and Savage should be listened to and non-oil companies should be encouraged to try their hand at the business. As a reminder, Savage suggested that oil processors and not rock handlers have been given too much importance to date in attempts to develop the resource. Welles maintained that the oil companies were afraid of oil shale because it could become a prime competitor for their valuable budget funds rather than a subsidiary or complementary operation. However, not many entities exist with the resources necessary to construct and operate an expensive energy facility. Walmart, notwithstanding its demonstrated ability to lower and stabilize gasoline prices in western Colorado, is probably not interested in building a shale oil plant. Petrochemical producers, automobile companies and energy transfer firms such as General Electric do seem to me to be possible candidates for shale oil sponsors. A partnership with an established engineering/procurement/construction firm would seem to be good synergism.

One of the things that is bothering me in today's rather unusual wartime situation is that there is no sense of urgency or sacrifice amongst the citizenry of our country. I am one of the folks who experienced Pearl Harbor as well as 9/11. I remember the early '40's when we knew there was a war on. Gasoline was rationed and controlled with a ration stamp permit system. We were mandated to observe a 35 mph speed limit no matter where we were going. The standard joke was that when the speed limit was lifted after the war, the car would not go faster than 36 mph. Cars were not just rationed, they were completely unavailable.

> "There's a Ford in your future, there's a Ford in your past, but the Ford you got now you'd better make it last."

We were restricted to one pair of new shoes a year. It seems like some of my grandkids get a pair a month. We all collected tin foil, rubber bands, and string for the war effort. Everyone my age eagerly bought war stamps in school which, when you had enough could be converted into war bonds. There was no chewing gum available. There was even "War Time", an accelerated daylight savings time designed to minimize electric power usage.

So it seems strange to me that we are constantly reminded that we are fighting a war against terrorism, but not really being asked to sacrifice. It would make lots more sense to me to have to give up something or things on a daily basis in order to commit myself better to our national cause. We, the people, have been actively searching for ways to help, giving too much blood to the Red Cross, collecting pennies for NYPD and firefighters, arguing over the political correctness of commendatory statues, etc. I know that the economists want us to keep spending money to avoid or mitigate recessions, but daily hardships, which can be blamed on bin Laden would enhance patriotism significantly. Frankly, a little more of the recession could currently be blamed on bin Laden if the spin doctors would do their job. What I'm personally most concerned about is that a worse recession will hit us as a result of a world-wide petroleum shortage. The similarities between 2003 and 1973 are scary. Again, development of a domestic oil shale supply industry could surely help reduce the overall effects of another Middle Eastern oil crisis.

Presidents and their administrations tried to get us to declare war on energy shortages in the 1970's. It didn't have too much effect. It maybe time to try that strategy again and get us all pulling together in a national effort to eliminate the problems of first, using so much energy and second, relying on unstable off-shore supplies for the energy we do use. In this vein, I received an e-mail dated October 4, 2001, purportedly from Representative Tom DeLay as follows:

"Dear Friend,

Defense of American freedom hinges on our national security, economic security, *and* our energy security. An overwhelming bipartisan majority in the United States House of Representatives knew this before the terrorist attacks on September 11 and passed the Securing America's Future Energy Act of 2001—the "SAFE" Act. The time has come for the Senate to act quickly on this important package. There are no excuses for waiting any longer. American families can't afford an energy policy that is increasingly dependent on foreign imports from a potentially unstable part of the world. Additional terrorist attacks or increased activities in the Middle East could ignite spiraling energy costs. In addition, each day we continue to depend heavily on oil imported from the Middle East, we are helping fund terrorists like Bin Laden who own interests there. Foreign terrorists are financed by oil money. Our dependence on foreign oil only strengthens regimes that have assisted the

international terror network. That's why we need to expand our
domestic energy supply to protect our national security. One key
step is opening a small 2,000-acre patch in the Arctic National
Wildlife Refuge (ANWR). This small amount of ANWR could
produce enough oil to meet the daily energy requirements of our
armed forces for an entire year. It could replace all the oil from Iraq
for the next fifty years! Opening ANWR would also create 700,000
jobs across the nation. That is twice as many new jobs as have been
lost since May 2001. In 1970, America imported 23% of its oil.
Last year, it was 57%. Unless we look for and develop new U.S.
reserves, reliance on foreign sources of oil will continue to rise. Here
are some staggering statistics:

-Rising energy costs drove us into our last three recessions.
-Soaring fuel prices last year cost the U.S. economy over $115
    billion.
-Over the next two years, high energy prices could cost farmers
    20% of their incomes.
-And over the next 20 years, America will:
-Consume 32% more energy
-Need 33% more petroleum
-Use 62% more natural gas
-Burn 22% more coal
-Require 45% more electricity.

We can't wait another day. The strength of our security and the
health of our economy rests on expanding our domestic energy
supply immediately. Failing to enhance our energy security only
strengthens the hands of people who want to harm us. Call your
Senators today, and encourage your friends and family across this
great nation to do the same. Tell them that you want to be energy
dependent America's security depends on it. Take care, Tom"

Perhaps a more modest approach to cranking up a renewal of oil
shale work would be more acceptable now. In my research, I came

across a reference to legislation passed in 1992 which provided for a 5 year program to further oil shale development. An excerpt from US Code: Title 42, Section 13412 is as follows:

**"Sec. 13412. Oil shale**

(a)  Program direction

The Secretary shall conduct a 5-year program, in accordance with sections 13541 and 13542 of this title, on oil shale extraction and conversion, including research and development on both eastern and western shales, as provided in this section.

(b)  Program goals

The goals of the program established under this section include—

> (1)  supporting the development of economically competi-
> tive and environmentally acceptable technologies to pro-
> duce domestic supplies of liquid fuels from oil shale;
> (2)  increasing knowledge of environmentally acceptable oil
> shale waste disposal technologies and practices;
> (3)  increasing knowledge of the chemistry and kinetics of oil
> shale retorting;
> (4)   increasing understanding of engineering issues
> concerning the design and scale-up of oil shale extraction
> and conversion technologies;
> (5)   improving techniques for oil shale mining systems; and
> (6)  providing for cooperation with universities and other
> private sector entities.

(c)  Eastern oil shale program

> (1)  As part of the program authorized by this section, the
> Secretary shall carry out a program on oil shale that

includes applied research, in cooperation with universities and the private sector, on eastern oil shale that may have the potential to decrease United States dependence on energy imports.

(2) As part of the program authorized by this subsection, the Secretary shall consider the potential benefits of including in that program applied research carried out in cooperation with universities and other private sector entities that are, as of October 24, 1992, engaged in research on eastern oil shale retorting and associated processes.

(3) The program carried out under this subsection shall be cost-shared with universities and the private sector to the maximum extent possible.

(d) Western oil shale program

As part of the program authorized by this section, the Secretary shall carry out a program on extracting oil from western oil shales that includes, if appropriate, establishment and utilization of at least one field testing center for the purpose of testing, evaluating, and developing improvements in oil shale technology at the field test level. In establishing such a center, the Secretary shall consider sites with existing oil shale mining and processing infrastructure and facilities. Sixty days prior to establishing any such field testing center, the Secretary shall submit a report to Congress on the center to be established.

(e) Authorization of appropriations

There are authorized to be appropriated to the Secretary for carrying out this section $5,250,000 for fiscal year 1993 and $6,000,000 for fiscal year 1994."

Through the good offices of our excellent Congressman, Scott

McInnis, I found that although this legislation was enacted, no funds were ever appropriated by Congress so the program was never implemented. Something like this seems appropriate at this point in time. A research effort which concentrates solely on processing mechanics and economics would be very useful. Other concerns such as the environment and socio-economic impacts should not be re-studied. All of the experience and data gained 20 years ago should be utilized rather than a new round of studies initiated. I urge you to consider all the data collected for Tract C-b as outlined in this book and then remember the same level of effort was performed for Tract C-a. Union Oil, Colony and other projects also developed vast amounts of useful data which do not need to be duplicated.

I unconditionally recommend that the research effort be directed by academia, say the Colorado School of Mines or some similar technical university level institution. My experience with oil shalers is that their claims for technology must be taken with a grain of salt. There are several test runs with the various technologies which could be analyzed and compared for efficacy and practicality. I seriously recommend that some of the RAND folks referenced in Chapter 6, or their disciples, be assigned a key role in the technological assessment. An international engineering firm with comparable but little direct shale experience would be logical as project administrator, coordinator and reporter. I recommend that the above-referenced 1992 legislation be resurrected or added to current energy plans as soon as possible.

Could this be the time for oil shale to be the "Treasure at the End of the Rainbow"? I believe that renewed efforts such as those described above should be undertaken to determine the proper size and development strategy for oil shale and thereby enhance our great nation's ability to thrive. I believe that it is the time to prove to Mr. Welles that man can be both prudent and a developer of United States shale oil resources.

I would refer you back to the discussions and conclusions of the Exxon "white paper" reproduced in Chapter 13 and Secretary of the Air Force Stetson's speech reproduced in Chapter 15. These discussions remain applicable today. In fact, you can almost always substitute the year 2000 every time that you see 1980 is these two presentations. We have been living with the problem of significant dependence on foreign oil for years and have done essentially nothing about it. I would also remind you that, according to Mssrs. Hook and Russell as outlined in Chapter 1, there are more barrels of oil in existence in oil shale around the world than in conventional petroleum deposits which could conceivably be used to supply the world for many years to come. An astounding potential 2000 trillion barrels have been identified around world-wide.

Some of today's politico-energy discussions center on increased conservation through various tactics such as increasing the allowable fuel consumption limits on our transportation systems. I firmly believe in conservation, but also believe that realization of this goal will only come about when it's not easier to buy and use big SUVs and houses than otherwise. In other words, the marketplace will dictate when it becomes best for me to give up my pickup and buy a small electric-powered van, for instance. The time that this will take will be significant if the past 20 years are any example. In the meantime, the US and indeed the world are using up all our supplies of petroleum. To get us more smoothly through the transition to a world without oil, the development of a shale oil industry seems very important to me. I think that the United States should continue its firmly established tradition of relying on native bountiful resources. As petroleum use becomes restricted to petrochemical feedstock and transportation, a reliable source of an acceptable substitute would be comforting. I love my country and the good life that I have been fortunate to enjoy. I sincerely wish for my grandchildren to be able to enjoy an equally fine life. A practical insurance policy to allow that could be an organized, methodical development of this known energy resource which would defer the use of significant quantities of oil and allow a smoother transition to a non-oil based economy in the future.

# ENDNOTES

[i] Hubbert, M. King, "Resources and Man, A Study and Recommendations" by the Committee on Resources and Man of the Division of Earth Sciences, National Academy of Sciences-National Research Council. W. H. Freeman and Company, San Francisco, 1969.

[ii] Deffeyes, Kenneth S., "Hubbert's Peak, The Impending World Oil Shortage" Princeton University Press. 2001

[iii] Published by Hubbert, M. King, 1962. *Energy Resources-A Report to the Committee on Natural Resources.* National Academy of Sciences-National Research Council Publication 100-D. Washington D.C.

[iv] Yergin, Daniel. "The Prize, The Epic Quest for Oil, Money and Power. Simon & Schuster, NY, 1991.

[v] Shale Oil Status Report, Prepared by RMOGA Committee on Oil Shale, Denver, Colorado, July, 1981

[vi] Horvarth, R.E. Editor, Selected Papers from the Energy Workshop: Industry

Perspectives on Pioneer Process Plants. RAND Corporation, Santa Monica CA, 1981

vii     Morrow, Edward W., Phillips, Kenneth E. and Myers, Christopher. Understanding Cost Growth and Performance Shortfalls in Pioneer Process Plants. RAND Corporation. Santa Monica, CA, 1981.

viii    Shale Country magazine, May, 1976.Shale Oil Economics: Big Dollars, Big Debt . . . Alys Novak.

ix      Shale Country April 1976. ERDA and FEA Look—Warily—at Shale. Alys Novak

x       The Synthetic Fuels Corp. Continues to Court Oil Shale. Shale Country Summer 1983. Anne Wasko

xi      "Exxon Mothballs Shale Project." Shale Country June/July 1982. pp1,2

xii     Denver Post November 29, 2001

xiii    Wasko, Anne. "Shale's Future: an Economic Matter." Shale Country December 1982

xiv     Stetson, John C., Secretary of the Air Force. Speech to SRI International, Menlo Park, California. March 8, 1979

xv      Huber, Peter, "Hard Green". Basic Books. New York. 1999 copyright (c) 1999 by Peter Huber. Reprinted by permission of Basic Books, a member of Persues Books, LLC.

xvi     Bahr, Howard. "The Year of Jubilo: a novel of the Civil War. Henry Holt and Company LLC, New York. 2000.

# GLOSSARY OF TECHNICAL TERMS, ACRONYMS AND ABBREVIATIONS

| | |
|---|---|
| < | less than |
| AGR | Above Ground Retort |
| AOSS | Area Oil Shale Supervisor |
| ARCO | Atlantic Richfield Company |
| ATV | All Terrain Vehicle |
| B/D | Barrels per Day |
| B/DOE | Barrels per Day Oil Equivalent |
| BLM | Bureau of Land Management |
| Btu | British Thermal Units |
| C-a | Colorado A shale oil tract |
| C-b | Colorado B shale oil tract |
| CITF | Cumulative Impact Task Force |
| CO | Carbon Monoxide |
| DDP | Detailed Development Plan |
| DDT | A pesticide |
| DOE | Department of Energy |
| EIS | Environmental Impact Statement |
| EMT | Emergency Medical Technician |

| EPA | Environmental Protection Agency |
|---|---|
| ERDA | Energy Research and Development Administration |
| FOSL | Fogey Obeying Speed Limit |
| gal/st | gallons per short ton |
| GNP | Gross National Product |
| gpm | gallons per minute |
| iiS | integrated in-situ |
| in situ | Latin for in place recovery |
| JBC | Joint Budget Committee |
| KV | Kilovolts |
| KW | Kilowatts |
| lagomorphs | Rabbits and Hares |
| lbs | Pounds |
| LPG | Liquefied Petroleum Gas |
| MIS | Modified In-situ |
| MMC | Multi-Mineral Corporation |
| MSHA | Mine Safety and Health Administration |
| NPDES | National Pollutant Discharge Elimination System |
| OPEC | Organization of Petroleum Exporting Countries |
| OSEAP | Oil Shale Environmental Advisory Panel |
| Oxy | Occidental Petroleum or Occidental Oil Shale Inc. |
| pH | A measure of the acidity of a system |
| Projecteer | A Project Engineer |
| PSD | Prevention of Significant Deterioration |
| R&D | Research and Development |
| RAMIS | Computerized air data base management system |
| RAND | Research and Development Corporation |
| SFC | Synthetic Fuels Corporation |
| SRI | Stanford Research Institute |
| STB | Staged Turbulent Bed |
| TOSCO | The Oil Shale Corporation |
| TPD | Tons per Day |
| TPSD | Tons per Stream Day |
| Ua | Utah A shale oil tract |
| Ub | Utah B shale oil tract |
| USers | Residents of the United States of America |
| V | Volts |

V/E            Ventilation/Escape
WRSP           White River Shale Project

# INDEX

# E

# F

# H

# I